Cambridge Elements

Elements in Decision Theory and Philosophy
edited by
Martin Peterson
Texas A&M University

EPISTEMIC GAME THEORY

Paolo Galeazzi
University of Bayreuth

Eric Pacuit
University of Maryland

Olivier Roy
University of Bayreuth

CAMBRIDGE
UNIVERSITY PRESS

CAMBRIDGE
UNIVERSITY PRESS

Shaftesbury Road, Cambridge CB2 8EA, United Kingdom

One Liberty Plaza, 20th Floor, New York, NY 10006, USA

477 Williamstown Road, Port Melbourne, VIC 3207, Australia

314–321, 3rd Floor, Plot 3, Splendor Forum, Jasola District Centre, New Delhi – 110025, India

Cambridge University Press is part of Cambridge University Press & Assessment, a department of the University of Cambridge.

We share the University's mission to contribute to society through the pursuit of education, learning and research at the highest international levels of excellence.

www.cambridge.org
Information on this title: www.cambridge.org/9781009618397

DOI: 10.1017/9781108756549

When citing this work, please include a reference to the DOI 10.1017/9781108756549

First published 2026

A catalogue record for this publication is available from the British Library

A Cataloging-in-Publication data record for this Element is available from the Library of Congress

ISBN 978-1-009-61839-7 Hardback
ISBN 978-1-108-70687-2 Paperback
ISSN 2517-4827 (online)
ISSN 2517-4819 (print)

Additional resources for this publication at www.cambridge.org/Pacuit

Epistemic Game Theory

Elements in Decision Theory and Philosophy

DOI: 10.1017/9781108756549
First published online: May 2026

Paolo Galeazzi
University of Bayreuth

Eric Pacuit
University of Maryland

Olivier Roy
University of Bayreuth

Author for correspondence: Eric Pacuit, epacuit@umd.edu

Abstract: This Element situates epistemic game theory at the intersection of decision theory, game theory, and interactive epistemology. It provides an overview and a critical assessment of some of the most classical results and contributions in the field: the epistemic characterization of Nash equilibrium, the epistemic interpretation of mixed actions, rationalizability in static games, and sub-game perfect equilibrium in dynamic games. The book furthermore discusses more recent contributions that highlight the importance of correlated beliefs in games, and as well as experimental and empirical findings on higher-order strategic reasoning.

Keywords: Epistemic game theory, decision theory, interactive epistemology, Foundations of game theory, higher-order interactive reasoning and strategic thinking

ISBNs: 9781009618397 (HB), 9781108706872 (PB), 9781108756549 (OC)
ISSNs: 2517-4827 (online), 2517-4819 (print)

Contents

1 What Is Epistemic Game Theory?

Imagine that you play the following game with a group of people. Each participant chooses an integer from 0 to 100 (meaning each participant selects one of the numbers 0, 1, 2, 3, . . ., up to and including 100). The goal is to pick a number that is two-thirds of the average of all the selected numbers. The person whose guess is closest to this target wins a 10 euro reward. In the event of a tie, the 10 euro prize is split between the winners. What number would you choose? If you're unfamiliar with this game, which is called *Guess the Average*,[1] stop reading for a minute and think carefully about what number you would pick and why you chose that number.

There are several strategic considerations that can influence your decision beyond simply picking a number at random:

- Avoid choosing 100 as your number – You will not win the full prize if you choose 100. If every player, including you, select 100, the prize of 10 euros is divided among all, yielding less than 10 euros per person. You would therefore have an incentive to choose 99 in that case. If at least one player picks a number less than 100, there are two possible scenarios: either one of the chosen numbers lower than 100 is the closest to the target, in which case you would have incentive not to choose 100, or 100 is still the closest to the target, in which case you would have incentive to choose 99. In both cases, 100 is not the best choice.

Of course, chances are slim that you are the *only* one among the participants who has come to this conclusion. Further strategic considerations require you to predict how the other participants are thinking about this game:

- Avoid choosing 99 as well – If most players are savvy enough to exclude 100 from their choices, then 99 becomes a high-risk number for similar reasons. If every participant, including yourself, chooses 99, then the prize is split, reducing the individual payout. Furthermore, if you choose 99 and others choose a smaller number, then you will not win the full prize, as explained earlier.

The reasoning does not stop at 99: If you predict that nearly all players will avoid choosing 99 or 100, then it makes sense to skip 98 as well. Following this line of reasoning, if everyone assumes others will think similarly,

[1] This game, also known as the "Guessing Game," goes back to the "Beauty Contest" from Keynes (1936). It has been extensively studied both empirically, for example, Nagel (1995), and theoretically, for example, Moulin (1986). See Nagel et al. (2017) for a history of this game.

97 becomes a poor choice too, and this pattern continues. How far will, or should, this reasoning go? We will look at some answers to these questions in Sections 4 and 7.

In thinking about what to choose in Guess the Average, we applied three main ideas, which also form the foundational pillars of Epistemic Game Theory (EGT), also known as the "Epistemic Program in Game Theory":

1. *Describing Situations of Strategic Interaction:* A situation of strategic interaction or, simply put, a game, is a scenario where multiple agents or players must independently decide their actions, knowing that the outcome depends on the decisions of the others too. In the aforementioned example, the players aim to guess a number closest to two-thirds of the average, aware that everyone else is doing the same. The success of their guess depends on the numbers chosen by all participants. Such situations are typically analyzed in *Game Theory*.
2. *Making Rational Decisions under Uncertainty:* When deciding what to do in a game, the players are uncertain about the outcomes of their decisions. In Guess the Average, each player is uncertain about the numbers others will choose, which complicates their decision. However, assuming that each player seeks to maximize their "material payoffs" (e.g., monetary gains), it becomes clear that choosing 100 is not optimal. Further, assuming that others recognize this reasoning, choosing 99 also becomes suboptimal, and so forth. This reasoning is grounded in *Decision Theory*, which explores rational choices under uncertainty.
3. *Representing the Players' Knowledge and Beliefs:* This concerns understanding each player's strategic uncertainty – how they think about the game and what they believe about how others are thinking about the game. Ruling out 99 assumes that others aim to maximize their gain and will similarly rule out 100. Excluding 98 involves additional assumptions that others believe everyone will rule out 99, and so on. This reasoning about the beliefs (and beliefs about the beliefs, and so on) of others, is studied in *Interactive Epistemology*.

Epistemic Game Theory integrates mathematical models and insights from these three areas–Game Theory, Decision Theory, and Interactive Epistemology (see Figure 1)–to explain what rational players would or should choose in strategic interactions. It emerged in the 1980s as a response to the so-called equilibrium refinement program in game theory (Van Damme, 1991). Epistemic Game Theory has since then developed a distinct perspective on rational decision-making and strategic thinking.

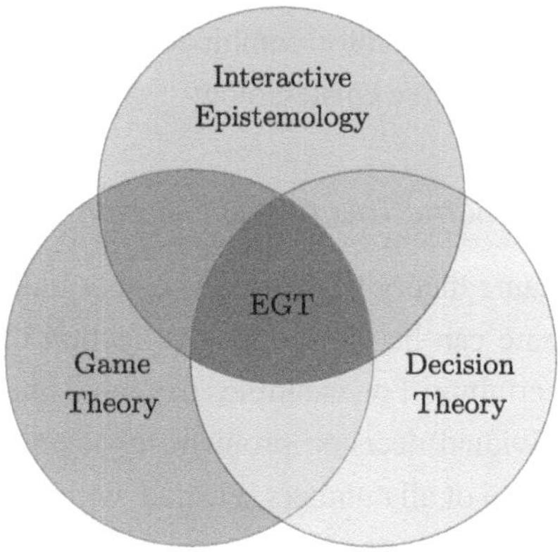

Figure 1 The three pillars of epistemic game theory (EGT): Decision Theory, Game Theory, and Interactive Epistemology

The remainder of this introductory section explains how Epistemic Game Theory builds on, yet differs from, Decision Theory, Game Theory, and Interactive Epistemology. We also provide an overview of its historical development. Throughout this discussion, we assume a basic familiarity with game theory and decision theory.[2] Readers who are new to these fields may wish to begin with Section 2 instead, which offers a concise review of key concepts and definitions.

1.1 Situating Epistemic Game Theory

Our perspective on epistemic game theory is nicely summarized by the following quote from Dekel and Sinischalchi's 2015 overview of the field:

> Epistemic game theory formalizes assumptions about *rationality* and *mutual beliefs* in a formal language, then studies their *behavioral implications in games*. [emphasis added]

As this quote suggests, Epistemic Game Theory combines two research projects: the study of mathematical models of strategic uncertainty (regarding the players' rationality, beliefs, and higher-order beliefs about one another) with the study of the "behavioral implications" of states of strategic uncertainty for rational players.[3] Both projects use standard models of rational

---

[2] For an introduction to game theory, see Osborne and Rubinstein (1994). For decision theory, see Peterson (2009) or Gilboa (2009).

[3] The study of mathematical models of strategic uncertainty is most prominent in, for example, Brandenburger (2014) and Perea (2012). The study of "behavioral implications" of states of rationality and of strategic uncertainty of the players dates back to the early contributions to the epistemic program (Spohn, 1982), and has been prominent ever since Pearce (1984),

decision-making under uncertainty combined with models of higher-order beliefs developed in interactive epistemology.

1.1.1 Game Theory and Decision Theory

From its inception, game theory has been intricately linked with decision theory.[4] Indeed, any game can be viewed as a collection of individual decision problems under uncertainty. For example, in Guess the Average, each participant faces an individual decision problem: choosing a number closest to two-thirds of the average of all numbers selected, without knowing the choices of the other participants. Of course, in this case, the winner(s) can only be determined after all numbers are submitted, which makes the players think about what the others choose, and why they do so. But there are variants of this game where this is not the case.

Imagine, for instance, that your task is to choose a number closest to two-thirds of the average of the numbers drawn at random by some devices, each selecting numbers based on a Gaussian distribution with some mean and standard deviation. Then your task is simply to try to estimate the parameters used by these devices, and select a number that approximates two-thirds of the expected average. In particular, in this case you do not have to think about what the devices expect you (or other devices) to choose.

As far as the decision rule is concerned, the two variants of the game are not fundamentally different. In each case, you choose the number that maximizes your expected (material) payoff, given your estimate of the average number chosen by the "others," whether they are humans or randomizing devices. The difference between these two lies in how you arrive at these estimates. In the original game, this process may involve considering the rationality and the beliefs of others.

These two cases illustrate the special role of decision-theoretic notions in epistemic game theory, namely to *explain* how rational decision-makers arrive at specific game-theoretic solutions. Historically, decision-theoretic representation theorems were used by game theorists as a justification for representing players' preferences using von Neumann-Morgenstern utilities (cf. Myerson, 1991). However, in epistemic game theory, the focus shifts toward using models of rational decision-making to explain why rational players might choose particular strategies in a game. As we will discuss throughout this Element,

Bernheim (1984), and Bacharach and Mongin (1994). See de Bruin (2010), Bonanno (2015), and Dekel and Siniscalchi (2015) for a discussion.

[4] Indeed, a fundamental representation theorem in decision theory appeared in the appendix of von Neumann and Morgenstern (1944), arguably the foundational text for game theory.

game-theoretic solution concepts identify specific profiles of choices as "solutions" to a game. Through studying the "behavioral implications" of certain states of strategic uncertainty for rational players, decision-theoretic tools provide part of the foundational basis for these solution concepts.[5]

1.1.2 Interactive Epistemology

Suppose that you are playing Guess the Average against a single opponent. If you both choose the same number, the 10 euro prize is equally divided. Therefore, it would be best for you to choose a number that is less than what your opponent selects. But this means that your best guess may be seen as depending on what you expect your opponent will guess, which, in turn, depends on what number your opponent expects that you will guess, and so on.[6] This pattern of 'mutual expectations' is common in many game-theoretic situations. Interactive epistemology studies mathematical models of knowledge or beliefs about others' knowledge or beliefs, which are essential to understand this type of back-and-forth reasoning.[7]

Interactive epistemology is essential for epistemic game theory because it provides the formal models used to represent states of strategic uncertainty and higher-order beliefs. As the opening quote in this section reveals, interactive epistemology is, in fact, often presented as an integral part of epistemic game theory. This is because many of the results in interactive epistemology have been obtained *by* (epistemic) game theorists, *with* game-theoretic applications in mind. However, it is useful to distinguish the two areas of research, even if the distinction is bound to remain fuzzy. For example, results such as the agreement theorems (Aumann, 1976) or the so-called Brandenburger-Kiesler

[5] It has been argued (Brandenburger, 2010) that one of the main motivations behind the development of the theory of zero-sum games in (von Neumann and Morgenstern, 1944) was to bypass decision-theoretic explanations, which might involve the threat of infinite regress. Indeed, it is clear that expected utility maximizers base their game choices on expectations about others' actions, which in turn depend on beliefs about what others believe, and so on. Morgenstern (1935) provides an early commentary on this phenomenon.

[6] But see Grosskopf and Nagel (2008) for a different perspective.

[7] The expression "Interactive Epistemology" was introduced to game theory in Aumann (1999) (see Aumann (2024) for an overview). We use it here in a broader sense that includes, for instance, multi-agent epistemic logic (van Benthem, 2011; van Ditmarsch et al., 2015; van Benthem and Klein, 2023), studies of "universal" models of beliefs (Böge and Eisele, 1979; Mertens and Zamir, 1985; Brandenburger and Dekel, 1993; Epstein and Wang, 1996; Heifetz and Samet, 1998a, 1998b; Battigalli and Siniscalchi, 1999; Moss and Viglizzo, 2004; Ganguli et al., 2016; Galeazzi and Marti, 2023; Fukuda, 2024), or the agreement theorem (Aumann, 1976) along with its extensions (Bacharach, 1985; Samet, 2010; Dominiak and Lefort, 2015; Samet, 2022). This use of "interactive epistemology" is less established than "decision theory" or "game theory," but we find it useful to highlight that all these studies focus on the mathematical modeling of higher-order knowledge and beliefs.

(BK) paradox (Brandenburger and Keisler, 2006) are of interest even outside of epistemic game theory, c.f. (Chant and Ernst, 2008; Lederman, 2015; Caie, 2016) for discussions of the agreement theorems, and (Abramsky and Zvesper, 2015) for the BK-paradox.

1.2 A Brief Historical Overview

The earliest contributions to the epistemic program include works by Armbruster and Böge (1979) and Böge and Eisele (1979), as well as Spohn (1982), but the importance of Harsanyi's (1967) work on Bayesian games[8] for the development of epistemic game theory, and of interactive epistemology more generally, is difficult to underestimate. Both Armbruster and Böge and Böge and Eisele explicitly acknowledged his influence. Harsanyi introduced type spaces – now essential for modeling players' higher-order beliefs (see Section 2) – initially to represent preference uncertainty in Bayesian games, but Armbruster and Böge and Böge and Eisele soon applied them to represent higher-order strategic uncertainty.

Despite the work of these precursors, it is fair to say that epistemic game theory really took momentum in the early to mid 1980s, as a response to the equilibrium refinement program (Selten, 1975; Myerson, 1978; Kreps and Wilson, 1982; Van Damme, 1984; McLennan, 1985; Van Damme, 1991). This research program started with the observation that Nash equilibrium does not always yield a unique or compelling solution in certain games. In response, many researchers sought to extend, refine, or modify existing equilibrium concepts to identify more plausible equilibria or exclude those deemed implausible, but these refined equilibrium concepts were often assessed on the basis of intuitive judgments about rational or plausible plays.

The earliest contributions to epistemic game theory attempted to move away from these intuitive criteria and instead base game-theoretic solutions on decision-theoretic foundations, supplemented by specific epistemic assumptions. Spohn (1982) is notable for having formulated this project early on, but the transition from the equilibrium refinement program to the epistemic program is best illustrated by the development of the concept of rationalizability by Bernheim (1984) and Pearce (1984), which we discuss in detail in Section 4. Even though the notion of rationalizable strategies captures the essence of what became known as the fundamental theorem of epistemic game theory, Pearce,

[8] Bayesian games represent strategic situations with incomplete information, where there is uncertainty about the structure of the game, such as the preferences or the actions available to the other players. Although not covered in this Element, we briefly discuss them in Section 2 and they come up in our discussion of the "purification theorem" in Section 3.

in particular, introduced the idea as an alternative solution concept, still very much in line with the equilibrium refinement program. Here is an excerpt from the abstract of Pearce's seminal (1984) paper:

> This paper explores the fundamental problem of what can be inferred about the outcome of a noncooperative game, from the rationality of the players and from the information they possess. The answer is summarized in a solution concept called rationalizability.

However, relatively soon after, understanding the epistemic foundations of game-theoretic solutions was routinely formulated as a research program of its own (e.g., Tan and da Costa Werlang (1988)), and references to the methodology of the equilibrium refinement program disappeared from most contributions to epistemic game theory (cf. Brandenburger (2010); Perea (2014b)).[9]

Epistemic game theory thus emerged from a need to move away from intuitive judgments about the plausibility of game solutions and instead provide them with more explicit foundations. Much of the work done today, and particularly the results surveyed in this Element, can still be understood in these terms. However, even though this foundational project initially focused on existing game-theoretic solutions, it is not limited to them. More recent developments have led to the emergence of new proposals for game-theoretic solutions based primarily on epistemic considerations. A notable example, discussed in Section 5, is the approach of ruling out beliefs in intrinsic correlations among the beliefs of others (Brandenburger and Friedenberg, 2008).

2 Background

This section is a brief introduction to the notation and concepts of decision theory, game theory, and interactive epistemology that will be used throughout the Element. It is not intended to be a comprehensive overview of these fields; readers seeking more detailed explanations and examples should consult an extended version of this section available in the Online Appendix, and the textbooks and papers referenced next. Nonetheless, this Element is largely self-contained, and no prior knowledge beyond what is covered here will be assumed.

2.1 Game Theory

A "game" is a playful term that refers to any multi-agent situation where rational players pursue goals while recognizing that outcomes depend on the

[9] We say that *the methodology* of the equilibrium refinement program has mostly disappeared because the refinements themselves are still very present in epistemic game theory. See the discussion in Section 8 for more details.

actions of other self-interested players, as illustrated by the Guess the Average game in Section 1. In this Element we only deal with so-called noncooperative game theory, where the players cannot make binding agreements with each other. See again classical textbooks like (Osborne and Rubinstein, 1994) for introductions to cooperative game theory.

For an interaction to be truly strategic, players must recognize that their choices' consequences depend on other self-interested agents' decisions. Strategic thinking about others only makes sense in these contexts. Consider the second variant of Guess the Average, where players estimate the mean of random drawings from a distribution rather than predicting others' choices. This represents a decision problem under uncertainty, not a game. Strategic thinking is what fundamentally distinguishes game theory from other decision frameworks.

Broadly, games fall into two categories: static and dynamic. In static games, players act simultaneously without prior knowledge of others' actions, as in Guess the Average. Dynamic games involve sequential play, where at least some information about the others' past actions is available to the players before making their own decisions. For simplicity, we begin our discussion with static games (dynamic games are discussed in Section 6).

2.1.1 Static Games

A static game is a tuple

$$(N, (A_i, \pi_i)_{i \in N})$$

consisting of the following components:

- a set N of players, with $|N| = n$ the number of players;
- a set A_i of actions for each player $i \in N$; and
- a payoff function $\pi_i : A \to \mathbb{R}$ for each player $i \in N$, where $A = \times_{i \in N} A_i$.

A game is said to be finite when both the set of players N and each action set A_i are finite. Unless specified otherwise, we assume that all the games discussed here are finite. Before proceeding, we clarify the following notation used for games.

- The notation $\times_{i \in N} A_i$ denotes the Cartesian product $A_1 \times A_2 \times \cdots \times A_n$.
- An action *profile* is a tuple of actions, one for each player. So, the set of all action profiles is the set $A = \times_{i \in N} A_i$.

		Player 2	
		f	g
Player 1	f	2, 1	0, 0
	g	0, 0	1, 2

Figure 2 An example of a static game, which is typically called the *Battle of the Sexes*

- When a is a profile, a_i denotes the i-th component of a, which is the action of player i.
- We also write $a_i, a'_i, a''_i, \ldots$ for arbitrary elements of A_i. The subscript i serves as a reminder that the action is from player i's set of actions.
- We write A_{-i} for the set of all tuples of actions for all players except i, that is, $A_{-i} = \times_{j \neq i} A_j = A_1 \times \cdots \times A_{i-1} \times A_{i+1} \times \cdots \times A_n$.
- When a is a profile, a_{-i} denotes the restriction of a to all the actions in a except i's action, that is, $a_{-i} = (a_1, \ldots, a_{i-1}, a_{i+1}, \ldots, a_n)$.
- We also write $a_{-i}, a'_{-i}, a''_{-i}, \ldots$ for arbitrary elements of A_{-i}.
- Given $a_{-i} \in A_{-i}$ and $a'_i \in A_i$, we write (a'_i, a_{-i}) for the action profile $(a_1, \ldots, a_{i-1}, a'_i, a_{i+1}, \ldots, a_n)$.

To illustrate the definition of a static game, consider a situation in which Ida and Joe agreed to meet to play either football[10] or golf. Joe's phone broke, so they cannot communicate to decide whether to meet at the football field or the golf course. They must independently choose whether to go to the football field or to the golf course. Both would rather play together, yet Ida prefers football (action f) while Joe prefers golf (action g). The static game representing this situation is shown in Figure 2.

Each player has two actions: going to the football field (f) and going to the golf course (g). By convention, the row player is Player 1 (Ida), and the column player is Player 2 (Joe). The action sets are $A_1 = A_2 = \{f, g\}$. If both decide to go to the football field, the outcome is that they play football, resulting in a higher payoff for Ida ($\pi_1(f, f) = 2, \pi_2(f, f) = 1$). If both go to the golf course, they go golfing, resulting in a higher payoff for Joe ($\pi_1(g, g) = 1, \pi_2(g, g) = 2$). If they do not meet, they simply go back home, resulting in the lowest payoff of 0 for both players ($\pi_1(f, g) = \pi_2(f, g) = \pi_1(g, f) = \pi_2(g, f) = 0$).

[10] Terminology note: by football, we mean soccer, ça va sans dire.

2.1.2 Game Solutions I: Nash Equilibrium

Nash equilibrium (Nash, 1950),[11] if not the most fundamental, is certainly the most famous solution concept in game theory. An action profile is a Nash equilibrium if no player has any incentive to unilaterally deviate from it. Formally, an action profile a is a *Nash equilibrium* (in pure actions) if

$$\pi_i(a_i, a_{-i}) \geq \pi_i(a_i', a_{-i})$$

for all $a_i' \in A_i$ and all $i \in N$. If the inequality is strict, the profile a is called a *strict Nash equilibrium*.

Consider the game depicted in Figure 2. To identify the Nash equilibria in this game, we examine the four possible action profiles: (f,f), (f,g), (g,f), and (g,g). It is clear that in the profiles (f,f) and (g,g), neither player has an incentive to change their action unilaterally. Hence, these profiles are Nash equilibria (in fact, they are strict Nash equilibria). In contrast, in the profiles (f,g) and (g,f), at least one player would benefit from changing their action (e.g., $\pi_2(f,f) > \pi_2(f,g)$). Therefore, these are not Nash equilibria.

The game in Figure 2 combines elements of coordination games (where players have a shared interest in coordinating their actions) and competitive games (where players have conflicting preferences about the outcomes). Consult the Online Appendix for more examples, including games with more than two players.

The Guess the Average game from Section 1 is an example of a static game with multiple actions and players. There, each of the n players chooses a number between 0 and 100, with those closest to $\frac{2}{3}$ of the average evenly splitting a 10 euro prize. Interestingly, depending on the number of players, the only Nash equilibria are 0 and 1. However, choosing these equilibrium strategies may not win the game. For instance, with more than two players, if the others choose large numbers, then picking 0 or 1 may not be a good choice.

2.1.3 Mixed Actions

Consider the static game shown in Figure 3. Examining this game, it is clear that for each action profile, at least one player has an incentive to unilaterally change their own action. Therefore, no profiles satisfy the Nash equilibrium

[11] See, for instance, Osborne and Rubinstein (1994) for a textbook presentations of the Nash equilibrium, and Holt and Roth (2004) for an historical overview of Nash's original paper.

		Player 2	
		h	t
Player 1	h	$1, -1$	$-1, 1$
	t	$-1, 1$	$1, -1$

Figure 3 The Matching Pennies game

definition provided in the previous section. An elegant solution to this problem is to allow players to choose probabilities over their actions.

A *mixed action* is a probability distribution over a player's own actions. We will discuss the interpretation of mixed actions in Section 3. Formally, a mixed action is an element of $\Delta(A_i)$, where $\Delta(A_i)$ is the probability simplex over A_i, that is, the set of all probability distributions over A_i. For an action $a_i \in A_i$ and a mixed action $\alpha_i \in \Delta(A_i)$, we write $\alpha_i(a_i)$ for the probability that player i plays a_i. Each $a_i \in A_i$ is called a *pure action* and can be viewed as a degenerate mixed action α_i, with $\alpha_i(a_i) = 1$.

Allowing players to choose mixed actions in a game generates a new game called the mixed extension. For a given game $(N, (A_i, \pi_i)_{i \in N})$, its mixed extension is another game

$$(N, (\Delta(A_i), \overline{\pi}_i)_{i \in N})$$

where

- N is the set of players;
- $\Delta(A_i)$ is the set of mixed actions of player i; and
- $\overline{\pi}_i : \times_{i \in N} \Delta(A_i) \to \mathbb{R}$ is the payoff function, defined as the expected payoff given action profile α:

  $$\overline{\pi}_i(\alpha) = E[\pi_i | \alpha] = \sum_{a \in A} \alpha(a) \pi_i(a)$$

  with $\alpha(a) = \prod_{i \in N} \alpha_i(a_i)$.

The definition of Nash equilibrium then naturally extends to profiles of mixed actions. A profile of mixed actions $\alpha = (\alpha_1, \ldots, \alpha_n)$ is a *mixed Nash equilibrium* of the game $(N, (A_i, \pi_i)_{i \in N})$ if it is a Nash equilibrium of the mixed extension. This means that a profile of mixed actions $\alpha = (\alpha_1, \ldots, \alpha_n)$ is a mixed Nash equilibrium if for every $i \in N$ and every $\alpha_i' \in \Delta(A_i)$,

$$\overline{\pi}_i(\alpha_i, \alpha_{-i}) \geq \overline{\pi}_i(\alpha_i', \alpha_{-i}).$$

If the inequality is strict, then the profile is called a *strict Nash equilibrium* in mixed actions. Only Nash equilibria in pure actions can be strict.

Returning to the Matching Pennies game from Figure 3, while there are no pure-action Nash equilibria, there is one mixed Nash equilibrium (α_1, α_2) where $\alpha_1(h) = \alpha_1(t) = 0.5$ and $\alpha_2(h) = \alpha_2(t) = 0.5$.[12] In fact, a fundamental result of game theory is that any finite game has at least one mixed Nash equilibrium (Nash, 1950).

2.1.4 Beliefs, Best Replies, and Dominance

In this section, we briefly introduce three fundamental concepts for epistemic game theory that will be explored in greater depth in later sections.

Beliefs: Given a game $(N, (A_i, \pi_i)_{i \in N})$, a *belief of player* i is a probability distribution β_i over the actions of the other players, that is, $\beta_i \in \Delta(A_{-i}) = \Delta(\times_{j \neq i} A_j)$.

Notice that there is a subtle but important difference between a profile of mixed actions $\alpha_{-i} \in \times_{j \neq i} \Delta(A_j)$ for all players except i and a belief $\beta_i \in \Delta(A_{-i})$ for player i. Each profile $\alpha_{-i} \in \times_{j \neq i} \Delta(A_j)$ can be associated with a belief $\beta_i \in \Delta(A_{-i})$, where for all $a_{-i} \in A_{-i}$, we set $\beta_i(a_{-i}) = \prod_{j \neq i} \alpha_j(a_j)$. Beliefs defined in this way assume that the actions of the other players are *independent* of one another. This is a significant assumption. Consider a game with three players: Ida, Joe, and Kai. Although all three players choose their actions independently of the other players, it may be reasonable for Ida, in some situations, to believe that Joe and Kai are acting in tandem and that their actions are hence not independent. We will come back to subtleties raised by such correlations in Section 5.

Best Reply: Consider a game $(N, (A_i, \pi_i)_{i \in N})$. A pure or mixed action $\alpha_i \in \Delta(A_i)$ is a *best reply* to some belief $\beta_i \in \Delta(A_{-i})$ for player i if

$$\alpha_i \in \arg\max_{\alpha'_i} E_{\beta_i}[\pi_i | \alpha'_i] = \arg\max_{\alpha'_i} \sum_{a'_i \in A_i} \sum_{a_{-i} \in A_{-i}} \alpha'_i(a'_i) \beta_i(a_{-i}) \pi_i(a'_i, a_{-i}).$$

A pure action $a_i \in A_i$ is said to be *justifiable* if there exists at least a belief β_i such that a_i is a best reply to β_i.

There is a relationship between mixed and pure best replies that is especially important in an epistemic setting. To introduce it, we need to define the notion of the support of a probability distribution. Given a mixed action α_i of player

12 Some games have both pure-action Nash equilibria and mixed-action Nash equilibria. For instance, in the game from Figure 2, in addition to the pure-action Nash equilibria (f, f) and (g, g), the mixed-action profile (α_1, α_2) where $\alpha_1(f) = \frac{2}{3}$, $\alpha_1(g) = \frac{1}{3}$ and $\alpha_2(f) = \frac{1}{3}$, $\alpha_2(g) = \frac{2}{3}$ is a mixed-action Nash equilibrium.

i, the *support of* α_i is the set of actions that are given nonzero probability by α_i. We denote the support of α_i by supp(α_i). It turns out that whenever a mixed action α_i is a best reply to a belief β_i for player i in a game $(N, (A_i, \pi_i)_{i \in N})$, each pure action in the support of α_i is itself a best reply to β_i. Formally, the following are equivalent:

1. $\alpha_i \in \arg\max_{\alpha'_i} E_{\beta_i}[\pi_i|\alpha'_i]$
2. $\text{supp}(\alpha_i) \subseteq \arg\max_{\alpha'_i} E_{\beta_i}[\pi_i|\alpha'_i]$.

Conceptually, this result means that, for any given probabilistic belief, the player has no need to randomize and play a mixed action: There is always a best reply in pure actions.

Consider again the Matching Pennies game depicted in Figure 3. If Ida, for instance, assigns a probability of 0.5 to Joe playing heads and a probability of 0.5 to Joe playing tails, then both heads (h) and tails (t) – as well as any mixed action of the two – are best replies for her. If instead, she assigns a probability of 0.7 to Joe playing tails and 0.3 to Joe playing heads, then the pure action tails (t) is the unique best reply for her.

Dominance: Suppose that $(N, (A_i, \pi_i)_{i \in N})$ is a game and $X \subseteq A_{-i}$ is a set of action profiles for all players except i. For $\alpha_i, \alpha'_i \in \Delta(A_i)$ we say that:

- α_i *strictly dominates* α'_i for i with respect to X provided that

$$\overline{\pi}_i(\alpha_i, a_{-i}) > \overline{\pi}_i(\alpha'_i, a_{-i}) \text{ for all } a_{-i} \in X.$$

- α_i *weakly dominates* α'_i for i with respect to X provided that

$$\overline{\pi}_i(\alpha_i, a_{-i}) \geq \overline{\pi}_i(\alpha'_i, a_{-i}) \text{ for all } a_{-i} \in X \text{ and}$$
$$\overline{\pi}_i(\alpha_i, a_{-i}) > \overline{\pi}_i(\alpha'_i, a_{-i}) \text{ for some } a_{-i} \in X.$$

There is an important connection between strict dominance and best replies. For any game $(N, (A_i, \pi_i)_{i \in N})$ and for any (mixed) action $\alpha_i \in \Delta(A_i)$ for player i, the following two statements are equivalent (Pearce, 1984):

1. a_i is strictly dominated with respect to A_{-i}.
2. a_i is not a best reply to any belief for i. That is, there is no $\beta_i \in \Delta(A_{-i})$ such that $a_i \in \arg\max_{a'_i} E_{\beta_i}[\pi_i|a'_i]$.

An analogous equivalence holds for actions that are weakly dominated, except that such actions are never best responses to full-support beliefs, that is, probability distributions that assign a positive probability to every profile of actions of the other players.

2.1.5 Game Solutions II: Iterated Dominance

Dominance reasoning can be iterated, leading to the concept of iterative elimination of strictly dominated actions. This process can be illustrated with the game depicted in Figure 4.

In this game, even though action r is not initially strictly dominated for Joe, it becomes strictly dominated after removing action m, which is strictly dominated for Ida. This reasoning by dominance, illustrated in Figure 5, goes as follows:

1. Action m is strictly dominated with respect to $A_{-1} = A_2 = \{l, r\}$ by the mixed action $\alpha_1(u) = \alpha_1(d) = 0.5$ for Ida, so m can be eliminated from Ida's actions.
2. In the reduced game with $A_1 = \{u, d\}$ and $A_2 = \{l, r\}$, the action r is strictly dominated by l for Joe, so it can be eliminated from Joe's actions.
3. In the reduced game with $A_1 = \{u, d\}$ and $A_2 = \{l\}$, the action d is strictly dominated by u, so it can be eliminated from Ida's actions.

After three iterations of eliminating strictly dominated actions, action u is the only action that survives for Ida and action l is the only action that survives for Joe. The iterated elimination of strictly dominated actions leads the players to the unique pure Nash equilibrium (u, l) of this game. This process is illustrated in Figure 5 with the strictly dominated actions crossed out at each step.

	l	r
u	4, 1	1, 0
m	2, 0	2, 1
d	1, 1	4, 0

G

Figure 4 A two-person game

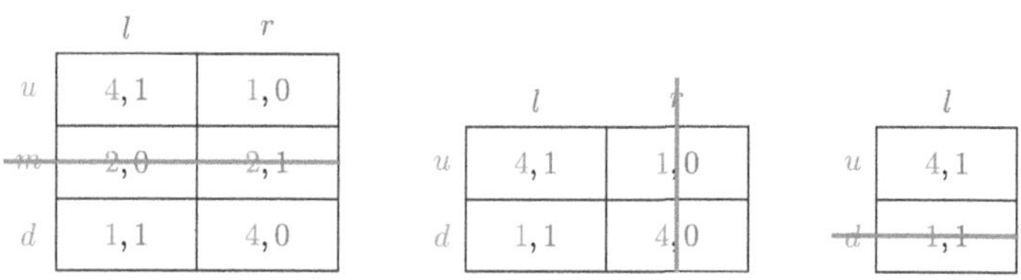

Step 1 Step 2 Step 3

Figure 5 Iteratively removing strictly dominated actions from the game in Figure 4

To formally define the aforementioned process, we need some notation. For a game $G = (N, (A_i, \pi_i)_{i \in N})$, let $X_i \subseteq A_i$, while as usual we write $X = \times_{i \in N} X_i$ and $X_{-i} = \times_{j \neq i} X_j$. Let $NSD_i(X)$ denote the set of player i's pure actions in X_i that are not strictly dominated with respect to X_{-i}:

$$NSD_i(X) = \{a_i \mid a_i \in X_i \text{ and } a_i \text{ is not strictly dominated with respect to } X_{-i}\}.$$

For instance, in the game depicted in Figure 4, we have $NSD_1(A) = \{u, d\}$ and $NSD_2(A) = \{l, r\}$. For any set X, let $NSD(X) = \times_{i \in N} NSD_i(X)$. Thus, in our running example, $NSD(A) = NSD_1(A) \times NSD_2(A) = \{u, d\} \times \{l, r\}$. This is the first step of elimination of strictly dominated actions.

The second step iteratively applies the NSD_i operator to $NSD(A)$:

$$NSD_1(NSD(A)) = \{u, d\} \text{ and } NSD_2(NSD(A)) = \{l\}.$$

We define $NSD^2(A)$ as follows:

$$NSD^2(A) = \times_{i \in N} NSD_i(NSD(A)).$$

Thus, in our example $NSD^2(A) = \{u, d\} \times \{l\}$. Continuing in this manner we have the following:

$$NSD^3(A) = NSD_1(NSD^2(A)) \times NSD_2(NSD^2(A)) = \{u\} \times \{l\}.$$

At this stage, the iterative process has reached a fixed point:

$$NSD^4(A) = NSD_1(NSD^3(A)) \times NSD_2(NSD^3(A)) = NSD^3(A).$$

In general, we define $NSD^n(A)$ recursively as follows: $NSD^1(A) = \times_{i \in N} NSD_i(A)$ and, for $k > 1$,

$$NSD^k(A) = \times_{i \in N} NSD_i(NSD^{k-1}(A)).$$

It can be shown that for any finite game, $NSD^k(\cdot)$ reaches a fixed point (see e.g. Chen et al. (2007); Apt (2011)), which we denote by $NSD^\infty(A)$. If an action profile a is in $NSD^\infty(A)$, we say that a survives iterated elimination of strictly dominated actions.

The iterative process defined earlier can also be applied to weak dominance defining an analogous operator $NWD_i(\cdot)$ for the set of actions that are not weakly dominated for player i. However, some complications arise when defining iterative elimination of weakly dominated actions. With strict dominance, the process that we defined earlier is *order-independent*, and it does not matter whether all strictly dominated actions are eliminated successively or all at once at each step of the process. This is no longer true when we iteratively remove weakly dominated actions. Consult Gilboa et al. (1990) and Apt (2007) for details, which are not discussed here, as they are beyond the scope of this Element. We briefly return to weak dominance in Section 4.

The reasoning process used in Guess the Average discussed in Section 1, where we first rule out 100, then 99, and so on, is a process of iterated elimination of strictly dominated actions. We can now state this more precisely. Consider a three-player Guess the Average game played by Ida (i), Joe (j), and Kai (k). As discussed in Section 1, playing 100 is never a best reply. Thus, as we saw earlier, this entails that 100 is strictly dominated.[13] For Ida, $NSD_i(A) = \{0, 1, 2, \ldots, 99\}$, and the same holds for Joe and Kai, where A is the set of all possible action profiles in the Guess the Average game.

Now, in the game where 100 is left out, 99 becomes the new highest number and, as we will see in Section 4, it is strictly dominated for similar reasons as earlier. Therefore, we get $NSD_i(NSD^1(A)) = \{0, 1, 2, \ldots, 98\}$, and the same applies for Joe and Kai. The process continues iteratively until we reach $NSD^{100}(A) = NSD_i(NSD^{99}(A)) \times NSD_j(NSD^{99}(A)) \times NSD_k(NSD^{99}(A)) = \{0\} \times \{0\} \times \{0\}$.[14]

2.2 Interactive Epistemology

Recall that we take interactive epistemology to be the study of models of beliefs and higher-order beliefs in interactive situations. These have been modeled in quite different ways. Here we start with the so-called probabilistic Kripke frames, closer to the types of model studied in logics of knowledge and belief, and then move to the so-called type spaces.

2.2.1 Probabilistic Kripke Frames

We start with models more familiar for those coming from the epistemic logic literature (e.g., Fagin et al. (1995)). A *probabilistic Kripke frame* is a tuple

$$(W, (b_i)_{i \in N})$$

where

- W is a set, the elements of which are called possible worlds, or states of the world. We assume that W is endowed with a σ-algebra[15] of events $\mathcal{W}$.

---

[13] Finding *what* action strictly dominates 100 is actually nontrivial. Although it is weakly dominated by any lower number, no pure action, that is no integer from 0 to 99, strictly dominates it. See Seel and Tsakas (2017) for the details.

[14] Note that if there are more than 3 players, then 1 is *not* strictly dominated by 0. In that case, both 0 and 1 survive iterative removal of strictly dominated actions. See Seel and Tsakas (2017) for the details.

[15] A σ-algebra $\mathcal{A}$ on a set X is a set of subsets of X that contains X and is closed under complement and countable unions.

	w_1	w_2	w_3	w_4	w_5	w_6
w_1	$3/10; 1/3$	$7/10; 0$	$0; 1/3$	$0; 0$	$0; 1/3$	$0; 0$
w_2	$3/10; 0$	$7/10; 3/4$	$0; 0$	$0; 1/4$	$0; 0$	$0; 0$
w_3	$0; 1/3$	$0; 0$	$1/5; 1/3$	$4/5; 0$	$0; 1/3$	$0; 0$
w_4	$0; 0$	$0; 3/4$	$1/5; 0$	$4/5; 1/4$	$0; 0$	$0; 0$
w_5	$0; 1/3$	$0; 0$	$0; 1/3$	$0; 0$	$0; 1/3$	$1; 0$
w_6	$0; 0$	$0; 3/4$	$0; 0$	$0; 1/4$	$0; 0$	$1; 0$

Figure 6 An example of a probabilistic Kripke frame

- $b_i : W \to \Delta(\mathcal{W})$ is a function that assigns a probability on $\mathcal{W}$ to each $w \in W$, representing player i's (posterior) beliefs at state w. We make the following assumption: For all $w \in W$ and $i \in N$, player i is *certain of their belief at* w. To formalize this, we need some notation: For $p \in \Delta(W)$, let $b_i^{-1}(p) = \{w' \in W : b_i(w') = p\}$. Then, i is certain of their belief at w provided that $b_i^{-1}(b_i(w))$ is measurable (i.e., an element of $\mathcal{W}$)[16] and

$$b_i(w)(b_i^{-1}(b_i(w))) = 1.$$

The elements of $\mathcal{W}$ are events, and the belief function b_i encodes the belief of player i at each possible world w. We can then represent a probabilistic Kripke frame as a matrix, such as the one depicted in Figure 6, which allows us to see at a glance how each player assigns probabilities to the possible worlds from the perspective of each world.

The matrix in Figure 6 represents a probabilistic Kripke frame $(W, (b_1, b_2))$ for two players, Ida and Joe, where $W = \{w_1, w_2, w_3, w_4, w_5, w_6\}$ and the σ-algebra $\mathcal{W}$ is the power set of W. In this matrix, row w_i gives the probability weights of Ida at world w_i as the first number of each cell, and the probability weights of Joe at world w_i as the second number of each cell. For instance, at world w_1 Ida assigns probability 3/10 to w_1 and 7/10 to w_2, while at world w_2 Joe gives probability 3/4 to w_2 and probability 1/4 to w_4.

In probabilistic Kripke frames, each world defines a hierarchy of beliefs for each player. Consider the world w_1 and the event $E = \{w_1, w_2, w_6\}$. At w_1, Ida is certain that E occurs since her belief function assigns it full probability:

$$b_1(w_1)(E) = 3/10 + 7/10 + 0 = 1.$$

Following standard terminology, we say that Ida believes E at state w_1. In general, when an agent gives an event E probability 1, we say that the agent is certain of E, or fully believes E, or simply believes E. Moreover, in this

[16] If $b_i^{-1}(b_i(w))$ is not measurable, then we assume $b_i^*(w)(b_i^{-1}(b_i(w))) = 1$, where b_i^* is the outer measure associated with b_i defined as $b_i^*(X) = \inf\{b_i(Y) : Y \in \mathcal{W} \text{ and } X \subseteq Y\}$. See Bjorndahl and Halpern (2017) for more details on this point.

model, the facts about players' beliefs are themselves events. The event that Ida believes E, for instance, is denoted by $\mathcal{B}_1(E)$ and is given by

$$\mathcal{B}_1(E) = \{w_1, w_2, w_5, w_6\}$$

because w_1, w_2, w_5 and w_6 are the worlds at which Ida gives probability 1 to the event E. Similarly, the event that Ida believes E^c (the complement or negation of E) is denoted by $\mathcal{B}_1(E^c)$, where

$$\mathcal{B}_1(E^c) = \{w_3, w_4\}.$$

From Figure 6, we can then infer that at w_1 Joe assigns probability $2/3$ to the event that Ida believes E and probability $1/3$ to the event that she believes E^c. Continuing in this way, the belief model generates the whole hierarchies of higher-order beliefs for both Ida and Joe at each state.

2.2.2 Type Spaces

The models most often used in game theory are the type spaces, which stem from the work of Harsanyi (1967) on Bayesian games. *A type space* is a structure

$$((T_i, b_i)_{i \in N})$$

where

- T_i is a set, the elements of which are called *types* of player i.
- $b_i : T_i \to \Delta(T_{-i})$ is the belief function for player i, where $T_{-i} = \times_{j \neq i} T_j$.

In a type space, the type of a player encapsulates their beliefs about the other players, as well as higher-order beliefs about what the other players believe, and so on. The belief function b_i specifies, for each type of player i, a probability distribution over the types of the other players, denoted as T_{-i}. A state in a type space is a profile of types $t \in T$, where $T = \times_{i \in N} T_i$, and an event is a set of states, $E \subseteq T$.

An example of a type space is given in Figure 7, where the types of Ida are $T_1 = \{t_1', t_1'', t_1'''\}$ and the types of Joe are $T_2 = \{t_2', t_2''\}$. Each row in Figure 7 corresponds to a type of Ida and specifies her probability distribution over Joe's types. The first number in each cell represents the probability that a given type of Ida assigns to the corresponding type of Joe. Similarly, each column represents a type of Joe and specifies his probability distribution over Ida's types. Joe's probability weights are the second numbers in each cell.

Notice therefore that, unlike in Figure 6, where the probability distributions of all agents at each world are specified in the rows, the probability distributions

	t_2'	t_2''
t_1'	$3/10; 1/3$	$7/10; 3/4$
t_1''	$1/5; 1/3$	$4/5; 1/4$
t_1'''	$0; 1/3$	$1; 0$

Figure 7 An example of a type space

of the types in Figure 7 are organized differently. In this figure, the probability distributions of Ida's types are specified by the rows, while the probability distributions of Joe's types are specified by the columns.

It is not hard to see that each type of a player generates a hierarchy of beliefs about the types of other players. For example, Joe's type t_2'' is certain that Ida's type is either t_1' or t_1'', which corresponds to the event $E = \{t_1', t_1''\} \times T_2$. Thus, the event that Joe is certain of E, denoted by $\mathcal{B}_2(E)$, is given by $T_1 \times \{t_2''\}$.[17] Similarly, Ida's type t_1''' is certain that Joe is certain that she is of type t_1' or t_1''. This reasoning extends recursively, generating whole hierarchies of beliefs.

Although there are some differences between type spaces and probabilistic Kripke frames, they are equivalent ways to represent the players' beliefs (Galeazzi and Lorini, 2016; Bjorndahl and Halpern, 2017; Bach and Perea, 2023). There is indeed a straightforward method to translate a type space into a Kripke frame that generates the same belief hierarchies, and vice versa. See the Online Appendix for a sketch of how this goes.

2.2.3 Game Models

In this section, we explain how to use probabilistic Kripke frames and type spaces to represent the players' beliefs in a game. Fix a game $G = (N, (A_i, \pi_i)_{i \in N})$. A *type space for G* is a structure

$$((A_i, T_i, b_i)_{i \in N})$$

where

- A_i is the set of actions of player i from G.
- T_i is the set of types of player i.
- $b_i : T_i \to \Delta(A_{-i} \times T_{-i})$ is the belief function of player i.

Player i is uncertain about both the actions and the types of the other players, and her belief is therefore a probability distribution over the Cartesian product $A_{-i} \times T_{-i}$. A state is given by a profile $(a_i, t_i)_{i \in N}$, which specifies both a type and an action for each player. An event is a set of states $E \subseteq A \times T$.

[17] In general, for an event $E \subseteq T$, the probability of E for type t_i of agent i is given by $b_i(t_i)(\{t_{-i} \in T_{-i} : (t_i, t_{-i}) \in E\})$.

A *probabilistic Kripke model for* G is a probabilistic Kripke frame augmented with a function assigning an action profile to each state:

$$(W, (b_i)_{i \in N}, \mathrm{s})$$

where

- $(W, (b_i)_{i \in N})$ is a probabilistic Kripke frame; and
- $\mathrm{s} : W \to A$, where $A = \times_{i \in N} A_i$.

We write $\mathrm{s}_i(w)$ for player i's action in $\mathrm{s}(w)$ and $\mathrm{s}_{-i}(w)$ for the profile of all actions in $\mathrm{s}(w)$ except for player i's action. For a profile of actions $a \in A$, we then write:

- $[a] = \{w : \mathrm{s}(w) = a\}$,
- $[a_{-i}] = \{w : \mathrm{s}_{-i}(w) = a_{-i}\}$,
- $[a_i] = \{w : \mathrm{s}_i(w) = a_i\}$.

So, for example, $[\mathrm{s}_i(w)]$ is the event that i chooses the action played in state w, and $b_i(w)([\mathrm{s}_{-i}(w)])$ is the probability in state w that i assigns to all other players playing their action in w.

There are two crucial assumptions that we make about probabilistic Kripke models for games throughout this Element:

1. Each of the aforementioned probabilities is well-defined (i.e., the aforementioned events are all measurable).
2. Players have no uncertainty about their own actions.[18]

2.3 Rationality, Beliefs, and Common Beliefs

Consider a game $(N, (A_i, \pi_i)_{i \in N})$. The game models presented in the last section describe the "context" of the game, specifying the players' (higher-order) beliefs. Both game models just introduced include the notion of event. In particular, three events play an important role in epistemic game theory:

- The event that player i is rational, denoted by R_i.
- The event that player i believes an event E, denoted by $\mathcal{B}_i(E)$.
- The event that there is common belief (among all players) in an event E, denoted by $\mathcal{CB}(E)$.

Acting Rationally: Consider a game $G = (N, (A_i, \pi_i)_{i \in N})$ and a probabilistic Kripke model $(W, (b_i)_{i \in N}, \mathrm{s})$ for G. Player i is *rational* at state w provided that

[18] In probabilistic Kripke models this means that for all $w \in W$, $b_i(w)([\mathrm{s}_i(w)]) = 1$. In the type spaces presented here the players' beliefs are not defined on their own actions.

$s_i(w)$ is a best reply[19] to this belief (recall the definition of best reply from Section 2.1.4). That is, agent i is rational at w provided that:

$$s_i(w) \in \arg\max_{a'_i} \sum_{a_{-i} \in A_{-i}} b_i(w)([a_{-i}])\pi_i(a'_i, a_{-i}).$$

The event that player i is rational is defined as $R_i = \{w : i \text{ is rational at } w\}$, and the event that all players are rational is $R = \bigcap_{i \in N} R_i$.

The definition of the event that all players are rational is similar in type spaces $((A_i, T_i, b_i)_{i \in N})$ for G. The key observation is that in a type space each type $t_i \in T_i$ generates a belief about the actions of the other players, denoted $b_i(t_i)(a_{-i})$, which is the marginal of $b_i(t_i)$ over A_{-i}. So an action a_i is rational for type t_i of player i whenever:

$$a_i \in \arg\max_{a'_i} \sum_{a_{-i} \in A_{-i}} b_i(t_i)(a_{-i})\pi_i(a'_i, a_{-i}).$$

Beliefs and Knowledge: Both probabilistic Kripke models and type spaces represent the players' beliefs as probabilities on events. In epistemic game theory it is customary to define probability-1 beliefs as *certainty*. That is, agent i believes an event E at state w when she assigns probability 1 to E at w. The event $\mathcal{B}_i(E)$ that agent i believes E is thus defined, in probabilistic Kripke models, as $\{w : b_i(w)(E) = 1\}$.

In Section 6, we will also be interested in what the players *know* about each other. "Knowledge" is a term of art here, and the reader accustomed to the rich discussions in epistemology should refrain from reading too much in this notion. For better or worse, epistemic game theory uses "knowledge" to denote *correct certainty*, that is, a belief with probability 1 that happens to be true. This use makes no assumption about how this belief happens to be correct, and whether the player has any kind of justification for it.[20]

This function satisfies the following properties.

---

[19] In this Element, we always assume that beliefs are probabilistic and best replying means maximizing expected utility. However, there are alternative models where these assumptions do not hold. See e.g. Gilboa and Schmeidler (1989); Schmeidler (1989); Hayashi (2008); Stoye (2011) for alternatives.

[20] There are, of course, well-known problems with identifying probability-1 beliefs with certainty and knowledge with correct certainty. There are, in particular, various alternative ways to define beliefs from a probability measure (Huber, 2009; Lin and Kelly, 2012; Leitgeb, 2017). Since our focus is on presenting existing work on the epistemic foundations of solution concepts, we do not discuss this literature here. See, however, Baltag et al. (2009) and Brandenburger et al. (2023) for alternative definitions of beliefs used to study dynamic games. We will come back to this, and to corresponding notions of correct certainty or knowledge in dynamic games, in Section 6.

- Players believe all "necessary truths," that is, events that are true everywhere in the frame or the type space: $\mathcal{B}_i(W) = W$. Observe that this also includes tautologies, which is one aspect of the so-called logical omniscience property.
- The players' beliefs are closed under necessary entailment. If an event entails another, then if a player believes the first, she also believes the second. Formally, we have that for all $E, F \subseteq W$, if $E \subseteq F$, then $\mathcal{B}_i(E) \subseteq \mathcal{B}_i(F)$. Observe, again, that this means, in particular, that the players' beliefs are closed under (classical) logical consequence, which is another facet of logical omniscience.
- The players' beliefs are also closed under intersection, which for our purposes can be seen as a form of conjunction: for all $E, F \subseteq W$, $\mathcal{B}_i(E \cap F) = \mathcal{B}_i(E) \cap \mathcal{B}_i(F)$.
- Each player's belief is consistent; that is, they never believe events that are necessarily false: $\mathcal{B}_i(\varnothing) = \varnothing$
- Finally, the players' beliefs are assumed to be fully introspective. If a player believes something, she believes that she believes it, and if a player does not believe something, she believes that she does not believe it. Formally, the first property, often called positive introspection, is defined as: For all $E \subseteq W$, $\mathcal{B}_i(E) \subseteq \mathcal{B}_i(\mathcal{B}_i(E))$. The second property, often called negative introspection, is defined similarly: For all $E \subseteq W$, $W \backslash \mathcal{B}_i(E) \subseteq \mathcal{B}_i(W \backslash \mathcal{B}_i(E))$, where $X \setminus Y = \{x : x \in X \text{ and } x \notin Y\}$.

Knowledge, understood minimally as correct certainty, is represented by a function $\mathcal{K}_i : \wp(W) \rightarrow \wp(W)$ satisfying all of the aforementioned conditions plus the additional assumption it is correct, that is, that $\mathcal{K}_i(E) \subseteq E$ for all events E.

Common Belief and Common Knowledge: As we will see in the next sections, the notions of common belief and common knowledge appear all the time in the epistemic foundations of game-theoretic solution concepts. Historically, the definition of common knowledge came first, going back at least to Lewis (1969). It has been introduced to game theory and interactive epistemology in Aumann (1976), and the definitions used there have become standard in the field.

There are multiple ways to mathematically define common knowledge and common belief (Barwise, 1988; Vanderschraaf and Sillari, 2023). Here we follow the standard practice in epistemic game theory and use the iterative definition. On that definition, common belief, and analogously common knowledge, can be seen as the limit of all (finite) iterations of "everybody believes that everybody believes that everybody believes . . .," with everybody believing a certain fact just meaning that each player believes that fact.[21] More precisely, *mutual* or *shared belief* in E, denoted $\mathcal{B}(E)$ (so the symbol $\mathcal{B}$ without an index for a specific agent), is defined as the event that all players believe E:

$$\mathcal{B}(E) = \bigcap_{i \in N} \mathcal{B}_i(E).$$

By iterating the mutual belief function we obtain arbitrarily high levels of mutual beliefs. Level 1 mutual beliefs in E is defined as $\mathcal{B}(E)$. Then for $n \geq 1$, mutual belief at level $n + 1$ is defined as the event that everyone believes that E is mutual belief up to level n:

$$\mathcal{B}^{n+1}(E) = \mathcal{B}(\mathcal{B}^n(E))$$

Common belief is then defined by taking the limit of all these finite levels of mutual knowledge

$$\mathcal{CB}(E) = \bigcap_{n \geq 1} \mathcal{B}^n(E)$$

So common belief means that the event E is not only mutually believed but also that this mutual belief is infinitely iterated; in other words, all players believe E, all players believe that all players believe E, and so forth, ad infinitum. Common knowledge $\mathcal{CK}$ is defined completely analogously. The only difference with common belief is that common knowledge inherits from the individual knowledge functions the property of being "true" or correct.

3 Nash Equilibrium and the Epistemic Interpretation of Mixed Actions

This section is focused on the epistemic foundation of Nash equilibrium in static games. As discussed in Section 2, every finite static game has a Nash equilibrium, though in some games, these equilibria are only in mixed actions. The epistemic analysis of Nash equilibrium is thus intertwined with the discussion of the various interpretations of mixed actions, and we will look at those in this section as well.

---

[21] The reader coming from philosophy should keep in mind that the iterative definitions of common knowledge and common belief do not necessarily correspond to the notions of common knowledge and common belief that are frequently used in the philosophy of collective action. See Paternotte (2011); Bozdag and Roy (2024) for a discussion and a comparison. Note also that the iterative definition is related to, but again not the same as, the one originally proposed by Lewis (1969). Cubitt and Sugden (2003) provides what is probably the most well-known formalization of Lewis' definition, but see also Vromen (2023) for a more recent take.

3.1 Correct Beliefs and Equilibrium Play

Recall that a Nash equilibrium is a profile of actions, possibly mixed, such that all players play their best reply given the actions chosen by the others. Nash equilibrium[22] and its numerous refinements have been extensively discussed in game theory (Aumann and Hart, 1992) and is arguably the solution concept that has been most used in philosophy (Lewis, 1969) and throughout the social sciences (Holt and Roth, 2004).

Under what conditions will rational players' choices be in equilibrium? In the early stages of epistemic game theory, researchers observed that equilibrium play requires players to have correct beliefs about – or even knowledge[23] – of the *choices* of others. Importantly, early characterizations of equilibrium play in static games do not require players to have beliefs about the rationality of others, nor do they depend on higher-order beliefs. The assumption that players hold correct beliefs about others' choices was already present in Armbruster and Böge (1979) and Spohn (1982), while Tan and da Costa Werlang (1988) and Aumann and Brandenburger (1995) emphasized that belief in – or knowledge of – the rationality of others is not central to the epistemic foundation of equilibrium play.

Before presenting the result, we illustrate the key ideas with an example. Figure 8 presents a game on the left and a type space associated with the game on the right. The game in Figure 8 is called the *chicken game*, where two young boys drive their cars to each other, and the first to swerve (d_i) is deemed the "chicken" and loses the game, very much like in the movie *Rebel Without a*

	d_2	h_2
d_1	0, 0	−1, 1
h_1	1, −1	−2, −2

	d'_2	d''_2	d'''_2	h'_2	h''_2
d'_1	$\frac{1}{2}; 0$	$0; \frac{1}{2}$	$0; 0$	$0; 0$	$\frac{1}{2}; 0$
d''_1	$0; 0$	$\frac{1}{2}; 0$	$0; \frac{1}{2}$	$\frac{1}{2}; \frac{1}{2}$	$0; 1$
h'_1	$0; 0$	$0; \frac{1}{2}$	$\frac{1}{2}; \frac{1}{2}$	$\frac{1}{2}; \frac{1}{2}$	$0; 0$
h''_1	$\frac{2}{3}; 1$	$0; 0$	$\frac{1}{3}; 0$	$0; 0$	$0; 0$
h'''_1	$1; 0$	$0; 0$	$0; 0$	$0; 0$	$0; 0$

Figure 8 A game (left) with an associated type space for the game (right)

22 Nash equilibrium is not the only equilibrium notion used in game theory. It has multiple refinements (Van Damme, 1983; Van Damme, 1991), for example, trembling hand equilibrium, subgame perfect equilibrium (Selten, 1975), or evolutionary stable strategies (Weibull, 1997). However, in this section we focus on Nash equilibrium and we often only write "equilibrium" instead of the full "Nash equilibrium."

23 Recall from Section 2 that "knowledge" and "belief" are technical terms in epistemic game theory. "Belief" refers to certainty, meaning belief with probability 1, while "knowledge" refers to correct certainty. Neither term is intended to capture any broader philosophical theory of knowledge or belief.

Cause with James Deen. This game is also well known in evolutionary game theory, where it is called the *hawk-dove game*. The labeling of the players' actions in Figure 8 comes from this second name. The evolutionary story behind this name is based on a population with two different groups of individuals: hawks and doves. Hawks are aggressive, while doves are not. When two doves meet, they do not fight, and nothing happens, resulting in a payoff of 0 for both players. When a hawk meets a dove, the hawk easily wins the confrontation, receiving a payoff of 1, while the dove gets −1. However, when two hawks meet, they both fight and harm each other, resulting in a payoff of −2 for each.

In the payoff matrix on the left, Ida is player 1 (as usual) with actions $A_1 = \{d_1, h_1\}$, and Joe is player 2 with actions $A_2 = \{d_2, h_2\}$. The type space on the right represents five different types for each player. Ida's types are $T_1 = \{d_1', d_1'', h_1', h_1'', h_1'''\}$, and Joe's types are $T_2 = \{d_2', d_2'', d_2''', h_2', h_2''\}$. Recall from Section 2 that a type space for a game associates a probability measure over the others' action-type pairs to each type of a player. For instance, each type of Ida is associated with a probability distribution over $A_2 \times T_2$. For compactness of representation, in Figure 8 we have directly associated the types of each player with an action. For example, d_2' represents Joe choosing action d_2 and being of type d_2'. A state is then defined just by a pair of types in this case, as each type is already associated with an action: The state (d_1', d_2'), for instance, represents the situation where Ida plays d_1 and believes with probability 0.5 that Joe is of type d_2' and plays d_2, and with probability 0.5 that Joe is of type h_2'' and plays h_2. On the other hand, at state (d_1', d_2') Joe plays d_2 and believes with probability 1 that Ida is of type h_1'' and plays h_1. In the formulation of Section 2, this would correspond in lengthier notation to the state (d_1, t_1', d_2, t_2'), where t_1' and t_2' are, respectively, the type of Ida and of Joe with the corresponding beliefs.

Notice that there are three Nash equilibria in this game, two in pure actions, (d_1, h_2) and (h_1, d_2), and one in mixed actions (α_1, α_2), where $\alpha_i(d_i) = \alpha_i(h_i) = 0.5$ for $i \in \{1,2\}$. From Section 2, we know that a Nash equilibrium in a two-player game is a profile of (possibly mixed) actions (α_1, α_2) such that Ida's action α_1 is a best reply to α_2 and Joe's action α_2 is a best reply to α_1.

Recall that, for an action a_i of player i, we denote by $[a_i]$ the event that player i plays action a_i. The following characterization is an immediate consequence of the definition of a Nash equilibrium.

Theorem 1: *Fix a static game G, a profile of actions $a = (a_1, \ldots, a_n)$ in G, and a type space (or, equivalently, a probabilistic Kripke model) for G. If all the players are rational at a state and correctly believe that a is being played at that state (and so a is indeed being played), then a is a Nash equilibrium.*

Notice that the aforementioned result only gives sufficient conditions for the Nash equilibrium: It may well be that Ida and Joe play an action at random and end up in equilibrium by luck. The argument for the result is straightforward. Rationality in games boils down to playing a best reply given one's belief about the choices of the others. If, moreover, the beliefs of all players turn out to be correct – that is, the players are actually playing what is believed – then each player is playing a best reply to the others' actions, and the profile of actions therefore constitutes a Nash equilibrium.

To better understand the result, the reader is invited to consider again the hawk-dove game and the associated type space from Figure 8. An instructive check shows for instance that at state (h_1'', d_2') both Ida and Joe are rational, have correct beliefs about the other, and the action profile (h_1, d_2) is indeed a Nash equilibrium.

3.1.1 Discussion

The correctness assumption is central to the characterization of equilibrium play, and this has led many scholars in the epistemic tradition to approach Nash equilibrium with skepticism. After all, how do players ever come to have *correct* beliefs of what others are choosing? This is a strong assumption, unlikely to hold in practice.

The characterization result we have outlined indeed explains equilibrium play as the outcome of rational players who know what others are doing. While this makes sense conceptually, it does little to explain how rational players could arrive at equilibrium in the more typical case where they are either uncertain or even mistaken about others' choices. Tan and da Costa Werlang (1988) and Bicchieri (1995) were among the first to articulate this concern,[24] and it has since become a common worry in epistemic game theory (Gintis, 2009; de Bruin, 2010; Perea, 2012).

There is no doubt that the correctness assumption substantially narrows the scope of the decision-theoretic explanation of equilibrium play, but concluding that the explanation is *too* narrow and therefore uninteresting might be too quick. Several authors, going back at least to Schelling (1960), have emphasized the importance of "converging expectations" for equilibrium play, especially when the players face an equilibrium selection problem. Correct mutual beliefs can be seen as a special case of such converging expectations. These authors have also identified various factors that can support such expectations, ranging from salience (Schelling, 1960; Harsanyi and Selten, 1988) and precedence (Lewis, 1969) to learning (Skyrms, 1990; Fudenberg and Levine,

[24] Although similar critical remarks can already be found in Bernheim (1984) and Pearce (1984).

1998), as well as collective perspectives (Bacharach, 1999) and social commitments (Gilbert, 1992). From an epistemic perspective, these factors suggest that assuming mutually correct beliefs is not always implausible. Put more simply, in many contexts, we seem to have enough evidence to make reliable guesses about others' choices – whether because we have interacted with them before, encountered similar situations, or because the game that is being playing is embedded in rich historical and cultural contexts (Lipman, 1995). How often this occurs is, of course, an empirical question.

However, even if we grant, for the sake of argument, that the epistemic perspective fails to provide a plausible explanation of equilibrium play, this does not justify blanket skepticism about Nash equilibrium as a solution concept. While the epistemic program is valuable, it is not the only possible foundation for Nash equilibrium. In fact, many of the works cited in the previous paragraph – especially those on learning in games (Skyrms, 1990; Fudenberg and Levine, 1998) – have successfully explained how agents converge to equilibrium play. The same is true of evolutionary approaches (see, e.g., (Weibull, 1997)). Arguably, such approaches may even offer stronger foundations for Nash equilibrium as a solution concept.

The widespread idea that game theory assumes common knowledge of rationality, perhaps in conjunction with the extensive use of equilibrium concepts in game-theoretic analysis, has led to the misconception that Nash equilibrium either *requires* common knowledge of rationality, or that common knowledge of rationality is *sufficient* for the players to play according to Nash equilibrium (see Bicchieri (1995) for a discussion of this point). Both of these claims are incorrect. If players hold mistaken beliefs but respond rationally, it is easy to construct examples where rationality is common knowledge, yet they fail to play a Nash equilibrium. Conversely, because choices in probabilistic Kripke models or type spaces are independent of the assignment of beliefs, one can also construct examples where players reach equilibrium despite rationality not being common knowledge. Common knowledge of rationality is neither necessary nor sufficient for Nash equilibrium play.

3.2 Mixed Actions and Their Interpretation

We have seen in Section 2 the example of Matching Pennies, a game that only has an equilibrium in mixed action, and we have already noted that the general existence of Nash equilibria in static games rests on such mixed actions. But what exactly are mixed actions? As it turns out, there has been much debate in game theory and in philosophy about the interpretation of mixed actions, and the epistemic perspective on games has much contributed to that discussion.

The traditional interpretation of a mixed action and mixed equilibria is in terms of genuine randomization. When a player adopts a mixed action, on that view, she is somehow randomizing, either in her head or using some device, and picks one of her pure actions with the probabilities specified by the mixed action. Sport provides numerous examples of "real" randomization in games, for example, in penalty kicks in football (Palacios-Huerta, 2014) or serving side in tennis (Ye et al., 2023).

This interpretation goes back to von Neumann and Morgenstern (1944), but there it was mainly focused on zero-sum games like Matching Pennies and has been criticized for a number of reasons. In such games, the players have an incentive to obfuscate their choices to their opponents, and this is naturally achieved by randomizing. This observation is less compelling, however, when players are not in direct competition, that is, for non-zero-sum games. It appears, furthermore, that we are generally not good at randomizing (Wagenaar, 1972), and some authors have argued that there can even be ethical reasons not to delegate one's choice to a randomizing device (Resnik, 1987; Rubinstein, 1991).[25]

The classical interpretation of mixed actions thus leaves game theory in somewhat of a conundrum. As we have seen, Nash equilibrium is central to game-theoretic analysis, not just in economics, but also in philosophy and in the social sciences. In many cases, however, equilibrium analysis requires considering mixed actions, but the classical interpretation of mixed actions only seems plausible in a rather limited class of cases. This basic difficulty with the classical interpretation of mixed actions has motivated the search for alternative interpretations, and epistemic game theorists have contributed to this by proposing the so-called *epistemic interpretation* of mixed actions. This interpretation has gained widespread adoption in epistemic game theory from the mid 1990s onward, roughly corresponding to the publication of Aumann and Brandenburger (1995). However, the epistemic interpretation has earlier historical roots in the so-called purification theorem (Harsanyi, 1973) and in the concept of correlated equilibrium (Aumann, 1974, 1987). It also appeared in early contributions to epistemic game theory (Armbruster and Böge, 1979; Tan and da Costa Werlang, 1988).

According to the epistemic interpretation, mixed actions represent *beliefs of the other players*. This can be decomposed into two points. The first is that players only choose pure actions. This claim by itself need not, however, lead to an epistemic interpretation of mixed actions just yet. As mentioned earlier, the

[25] See also Icard (2021); Zollman (2022) for different perspectives on this.

steady-state view from evolutionary game theory is one popular, non-epistemic interpretation of a mixed-action Nash equilibrium. Under that interpretation, mixed actions reflect distributions of pure actions in large populations of players (Weibull, 1997), and mixed actions are thus not interpreted as subjective probabilities, while players do not randomize.

The second point, which is arguably the core of the epistemic interpretation, is that a mixed action of a player represents uncertainty *of others* about what that player will do. Spohn (1982) provides a remarkably early version of this result. Its modern formulation, for two players, is due to Tan and da Costa Werlang (1988), who also credit Armbruster and Böge (1979) for the result, and the generalization to three or more players is due to Aumann and Brandenburger (1995).[26]

To illustrate how this works, let us first revisit Theorem 1, which applies to both pure and mixed actions. In a two-player game, a first-order belief of Ida is a belief on Joe's actions, and a second-order belief of Ida is a belief on Joe's beliefs on her actions, and so on. In the following, we denote a typical first-order belief of Ida by $\beta_1 \in \Delta(A_2)$, or more explicitly by stating the specific distribution over Joe's actions; for example, $(0.3d_2, 0.7h_2)$ and $(1h_2)$ are possible first-order beliefs of Ida in the hawk-dove game. With this in hand, Theorem 1 can be turned into a result essentially establishing the epistemic interpretation of mixed actions in two-player games.

Theorem 2 (*Aumann and Brandenburger, 1995*)**:** *Let G be a two-player static game, and consider a type space (or equivalently, a probabilistic Kripke model) for it. If, at a given state in the type space, there is mutual belief in rationality and mutual belief of the first-order beliefs, then the pair of first-order beliefs forms a Nash equilibrium of G.*

If one adopts such an epistemic perspective on mixed actions, then the interpretation of Nash equilibria shifts from equilibria in actions to what has been termed *equilibria in beliefs*. We will revisit this terminology in the discussion next. See the Online Appendix for illustrations of this result relative to the game and the type space in Figure 8.

Extending Theorem 2 to three or more players is less trivial than one might expect (see Tan and da Costa Werlang (1988); Bicchieri (1995) for an early

[26] Note, again, that the epistemic interpretation is not the only interpretation of mixed actions that view the probabilities as subjective credences. The so-called purification theorem from Harsanyi (1973) and Aumann's (1974) interpretation using exogenous signals are two other classical examples.

statement to that effect). Several solution exists,[27] but here we follow the two-steps approach proposed by Aumann and (Brandenburger, 1995, Theorem B).

The first key idea is to endogenize the role that is played by the commonly known probability on the players' signals in Aumann (1974). This is typically done by assuming that the players have common prior beliefs. The players' posteriors are then computed by conditioning this common prior on the players' private information, which typically includes the player's type and choice of pure action.

The second key idea is to make sense of the notion of *the* subjective belief of the other players about a player's choice of pure action. In two-player games, this is not a problem since there is only one opponent. A player's mixed action can be directly read off from her co-player's belief. However, with more than two players, nothing in principle prevents different players from having different beliefs about what some others will do, even if they have common prior beliefs. Thus, with more than two players, the notion of *the* belief of the other players does not make sense. The assumption that posterior probabilities are commonly known, under a common prior, circumvents this difficulty, since by the so-called agreement theorem (Aumann, 1976) these two assumptions together entail that the posteriors are the same.

We can now state the result formally.

Theorem 3: *Let G be a static game and w a state in either a probabilistic Kripke model or a type space for G. Suppose that*

1. *there is a common prior on the set of states such that the choices of all the players are independent,*
2. *all players are rational at w,*
3. *all beliefs, b_i, for player i assign probability 1 to the other players being rational at w, and*
4. *the players' posterior beliefs at w about the other players are common knowledge (i.e., the event that each player i's belief is $b_i(w)$ is common knowledge).*

Then all the players' induced beliefs about the actions of others are the same and, furthermore, these beliefs correspond to a mixed action Nash equilibrium.

[27] See e.g. Perea (2007b), Tan and da Costa Werlang (1988), Bach and Tsakas (2014), Barelli (2009), and Brandenburger and Dekel (1987) for similar results or alternative characterizations.

If we lift the assumption that the common prior beliefs are independent, then the beliefs about each player induced by the other players' belief form a mixed-action correlated equilibrium. If we lift the common prior assumption altogether, but instead assume that rationality is common knowledge, we obtain that the beliefs about each player induced by the other players' beliefs can be seen as *rationalizable* mixed actions (Brandenburger and Dekel, 1987), which we will discuss in more detail in the next section.

It is important to emphasize again that neither this result nor Theorem 2 are characterizations of equilibrium *play*. This is so because, as we have seen in Section 2, when a mixed action is a best reply, any action in the support of that mixed action is a best reply to the (possibly mixed) action played by the others. Thus, in any state satisfying the assumptions of the aforementioned theorems, any action for a player i in the support of the other players' beliefs about her choice is a best reply to i's belief about what the others will do.

3.2.1 Discussion

Just like the epistemic characterization of equilibrium play, the epistemic interpretation of mixed actions rests on rather strong assumptions, namely common prior belief and common knowledge of posteriors. Common prior belief is a rather standard assumption in economic analysis, but it is not uncontroversial. Morris (1995) presents a thorough overview of the arguments in favor of that assumption, and notably concludes that none of them are entirely convincing. If one does not assume a common prior, many, and perhaps *too* many differences in behavior could be explained by differences in priors. The assumption of common knowledge of posterior beliefs is used to eliminate possible differences in those posteriors beliefs and, by the same token, to be able to speak of "the" (unique) belief of the others about one's action. Without this, the epistemic interpretation of mixed actions will, in most cases, not be well defined for more than two players.

Although the terminology is relatively common, in this section we have tried to use "equilibrium of beliefs" (Volij, 1996) as little as possible. The analogy with equilibria in pure or mixed actions is indeed unclear. It makes little sense to say that the beliefs are mutual best responses to each other. This requires being able to talk about the players' incentive, or lack thereof, to change their beliefs. Of course, players might change their beliefs about each other as a result of observing certain signals or, in dynamic games, previous choices. But these belief changes are driven by new evidence or new information, not by strategically adjusting their expectations, that is, deciding what to believe, because this makes them better off.

The epistemic interpretation of mixed actions loses some of its appeal when we move to games with more than two players. The main conceptual difficulty is that the interpretation requires us to speak of "the" (unique) expectation of others about one's own choices. With more than two co-players, these expectations need to be the same. Absent an explanation of how this might happen, the epistemic interpretation seems to be applicable only in exceptional circumstances where one's co-players have identical beliefs about what one might choose.

As with Nash equilibrium, the conceptual difficulties faced by the epistemic interpretation do not justify general skepticism about the use of mixed actions in game theory. First, the problem does not arise in two-player games, which include many of the games analyzed in philosophy. Second, recall that the search for an alternative interpretation of mixed actions was motivated by the observation that players are often either unable or unwilling to randomize. We briefly noted that the epistemic approach is not the only way to interpret mixed actions while avoiding the assumption that players genuinely randomize. The steady-state interpretation, for example, also avoids this assumption.

4 The Fundamental Theorem of Epistemic Game Theory

In Section 2, we introduced the concept of (iterated) elimination of strictly dominated actions. The fundamental theorem of epistemic game theory establishes a link between this elimination procedure and the epistemic conditions that lead to it.

4.1 Strong Dominance and Rationality

Recall from Section 2 that an action a_i is justifiable if there exists a belief $\beta_i \in \Delta(\times_{j \neq i} A_j)$ such that a_i is a best reply to β_i. Playing a best reply to one's own belief is equivalent to choosing an action that maximizes subjective expected utility, making a *rational* decision under uncertainty. It follows immediately that a rational player never chooses a non-justifiable action.

Given a static game $(N, (A_i, \pi_i)_{i \in N})$, we can define the set of best replies of agent i given belief β_i by a correspondence $br_i : \Delta(A_{-i}) \rightrightarrows A_i$ such that[28]

$$br_i(\beta_i) = A_i \cap \arg\max_{\alpha_i} E_{\beta_i}\left[\pi_i | \alpha_i\right]$$

28 The terminology "correspondence" and notation $f : X \rightrightarrows Y$ mean that f is a function from X to subsets of Y.

	l	r
u	4, 1	1, 0
m	2, 0	2, 1
d	1, 1	4, 0

G

Figure 9 A two-player game

Thus, the set of justifiable actions of player i is defined as

$$br_i(\Delta(A_{-i})) = \{a_i \in A_i : \exists \beta_i \in \Delta(A_{-i}), a_i \in br_i(\beta_i)\}.$$

These are the only actions that a rational player would choose. Notice that, as defined earlier, the set of justifiable actions consists only of pure actions. This is justified (pun unintended) by the equivalence results between pure and mixed best replies presented in Section 2.1. Furthermore, this choice is motivated by the fact that in non-repeated static games, each player selects an action only once. As a result, the only observable aspect of a player's behavior in the game is a single pure action.

The first step toward the fundamental theorem of epistemic game theory is the following result, proven by Pearce (1984): for all $i \in N$,

$$br_i(\Delta(A_{-i})) = NSD_i(A).$$

That is, the set of justifiable actions is exactly the set of actions that are not strictly dominated. Despite its simplicity, this is a profound result. While the two sides of this equivalence may appear similar, they capture fundamentally different ideas. On one side, we have an elimination procedure that removes certain actions; on the other, we have the concept of a best reply based on a player's beliefs about their opponents' actions.

To better understand this result, recall the game from Figure 4, which we reproduce here in Figure 9 for convenience. Regarding strictly dominated actions, we saw in Section 2.1 that initially Joe has no strictly dominated actions, while for Ida action m is strictly dominated by the mixed action α_1, where $\alpha_1(u) = \alpha_1(d) = 0.5$.

Determining the players' best replies for every possible belief is not immediate. Since Joe initially has no strictly dominated actions, we focus on Ida and illustrate her expected payoffs as a function of her beliefs in Figure 10. The horizontal axis represents the probability that Ida assigns to Joe playing action r. This probability ranges from 0, when she is certain that Joe chooses l, to 1, when she is certain that Joe chooses r. Since Ida's belief in this case

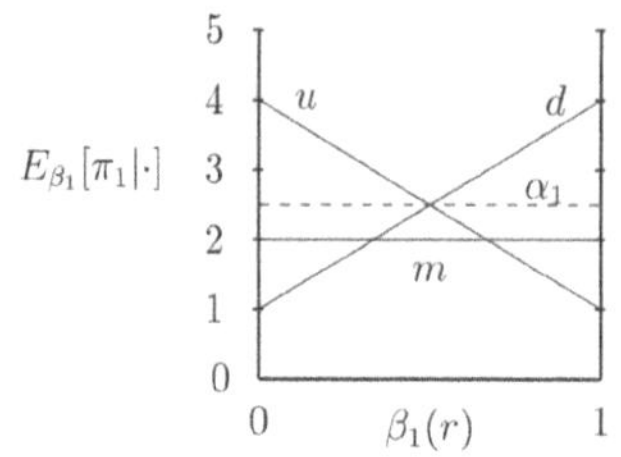

Figure 10 Ida's expected payoffs for the game in Figure 9

is fully determined by the probability she assigns to r, we can express it as $\beta_1(l) = 1 - \beta_1(r)$.

The vertical axis represents the expected payoff to Ida for any of her possible actions. For instance, her expected payoff from playing u while believing that Joe plays l, $\beta_1(r) = 0$, is 4, whereas her expected payoff from u while believing $\beta_1(r) = \beta_1(l) = 0.5$ is 2.5. Notice also that the dashed horizontal line represents the expected payoff of the mixed action α_1, where $\alpha_1(u) = \alpha_1(d) = 0.5$.

This visualization makes two important points clear. First, action m, which yields a fixed payoff of 2, is strictly dominated by the mixed action α_1, which yields an expected payoff of 2.5 against both l and r. Second, there is no belief Ida could hold – that is, no point on the horizontal axis between 0 and 1 – for which m is a best reply. To the left of $\beta_1(r) = 0.5$, action u maximizes her expected payoff and is therefore her best reply. To the right of $\beta_1(r) = 0.5$, action d is her best reply. At exactly $\beta_1(r) = 0.5$, both u and d are best replies.

4.2 Believing in Rationality

As discussed in Section 1, epistemic game theory focuses on players' strategic reasoning. In Section 2, we introduced type spaces and probabilistic Kripke frames as models for representing these beliefs. These models are used to represent higher-order beliefs. However, as we saw in Section 3, when analyzing strategic reasoning in games the most relevant beliefs concern the *choices* of other players and whether these choices are *rational* – meaning they are best replies to others' beliefs.

Figure 11 represents one possible type space for the game in Figure 9. As in Section 3, we include each player's chosen action in the type notation for compactness.

In this type space, the event $[l] \subseteq A \times T$ that Joe plays l is then given by:

$$[l] = \{(u_{i_1}, l_{j_1}), (m_{i_1}, l_{j_1}), (d_{i_1}, l_{j_1}), (u_{i_1}, l_{j_2}), (u_{i_1}, l_{j_2}), (u_{i_1}, l_{j_2})\}.$$

From Section 2, we know that the event where Joe believes (with probability 1) that Ida plays u is denoted by $\mathcal{B}_j([u])$. Thus, this event is given by:

	l_{j_1}	r_{j_1}	l_{j_2}	r_{j_2}
u_{i_1}	1/2; 1	0; 1	0; 1/3	1/2; 1/3
m_{i_1}	1/2; 0	0; 0	0; 2/3	1/2; 2/3
d_{i_1}	1/2; 0	0; 0	0; 0	1/2; 0

Figure 11 A type space for the game in Figure 9

$$\mathcal{B}_j([u]) = \{(u_{i_1}, l_{j_1}), (m_{i_1}, l_{j_1}), (d_{i_1}, l_{j_1}), (u_{i_1}, r_{j_1}), (u_{i_1}, r_{j_1}), (u_{i_1}, r_{j_1})\}.$$

Ida's and Joe's beliefs about each other's rationality are modeled similarly. Rationality in type spaces or probabilistic Kripke models corresponds to an event where a player's action is a best reply to her beliefs. For instance, if Ida believes that Joe plays l and she plays u, then she is rational, as u is a best reply to l. If instead she believes that Joe plays r and she plays u, she is not rational, as u is not a best reply to r. We denote the event that Ida is rational by R_i, and likewise R_j is the event that Joe is rational. The event R that both players are rational is the intersection of the events that each player is rational, $R = R_i \cap R_j$. In the example of Figure 11, we then have that

$$R = \{(u_{i_1}, l_{j_1}), (u_{i_1}, r_{j_2}), (d_{i_1}, l_{j_1}), (d_{i_1}, r_{j_2})\}$$

In a strategic setting, what is the behavioral implication of the event R, which states that the players are rational? We already know the answer from the results of Pearce (1984) presented earlier: Rational players will only choose actions that are not strictly dominated. But what if we consider agents who are slightly more sophisticated – not only rational themselves but also believing that others are rational too? What are the behavioral implications for the game? In other words, how would such agents play?

Consider Joe's position in the game shown in Figure 9. We know that if Joe is rational, he only chooses actions that are not strictly dominated. Furthermore, if Joe believes that Ida is rational, then by definition, he believes that she also avoids strictly dominated actions. This mean that if Joe believes Ida is rational, he believes that she plays either u or d, but not m. Given this belief, Joe expects to receive a payoff of 1 from choosing l and a payoff of 0 from choosing r. Consequently, he should choose l.

In our case, the event that Ida believes that Joe is rational includes all the states of our type space, $\mathcal{B}_i(R_j) = A \times T$, as all of Ida's types give probability 1 to the event that Joe is rational R_j. The event that Joe believes that Ida is rational is

$$\mathcal{B}_j(R_i) = \{(u_{i_1}, l_{j_1}), (m_{i_1}, l_{j_1}), (d_{i_1}, l_{j_1}), (u_{i_1}, r_{j_1}), (m_{i_1}, r_{j_1}), (d_{i_1}, r_{j_1})\}.$$

The event that the players believe that the other is rational is $\mathcal{B}(R) = \mathcal{B}_i(R) \cap \mathcal{B}_j(R)$, and the event that the players are rational and believe that the other is rational is $R \cap \mathcal{B}(R) = \{(u_{i_1}, l_{j_1}), (d_{i_1}, l_{j_1})\}$. At this point, only types of Joe that play l survive. Of course, this closely resembles the process of iterated elimination of strictly dominated actions. This similarity is no coincidence, as we will see in the next section, where we finally reach our goal, the fundamental theorem of epistemic game theory.

4.3 Rationalizability

To formally present the fundamental theorem, we need one additional component, the rationalization operator ρ. Recall that in a static game, the set A consists of all possible action profiles for the players. Let $\mathcal{A}$ denote the collection of all subsets of A that can be expressed as Cartesian products of individual players' action sets. The rationalization operator ρ takes elements of $\mathcal{A}$ as input and outputs elements of $\mathcal{A}$ as well. Intuitively, for a game itself, or any of its subgames, described by a subset of action profiles $A' \subseteq A$ – that is, an element of $\mathcal{A}$ – the operator ρ returns the set of action profiles in A that are justifiable based on some belief over the set of profiles in A'. More precisely, the operator $\rho : \mathcal{A} \to \mathcal{A}$ is defined in terms of individual rationalizability operators ρ_i as follows:

$$\rho_i(A'_{-i}) = br_i(\Delta(A'_{-i})) = \{a_i \in A_i : \exists \beta_i \in \Delta(A'_{-i}), a_i \in br_i(\beta_i)\}.$$

We then take $A'_{-i} = \times_{j \neq i} A'_j$, as usual, and $\rho(A')$ is defined as $\times_{i \in N} \rho_i(A'_{-i})$. Therefore, $\rho_i(A'_{-i})$ is the set of best replies of player i when her beliefs about the opponents are limited to the opponents' actions in A'_{-i}.

It is the iterated application of the operator ρ that is particularly relevant for our purposes here. To understand the mechanics of such iterations of ρ, let us again consider the game in Figure 9. In the first round, starting with A, we know from the result by Pearce (1984) discussed in Section 4.1 that

$$\rho(A) = br_i(\Delta(A_j)) \times br_j(\Delta(A_i)) = NSD(A).$$

Therefore, $\rho(\{u, m, d\} \times \{l, r\}) = \{u, d\} \times \{l, r\}$. The second iteration of the rationalization operator then corresponds to the application of ρ to this outcome. As we have argued previously, r is no longer rational for Joe once m is not available, and hence we have

$$\rho^2(A) = \rho(\rho(A)) = \rho(\{u, d\} \times \{l, r\}) = \{u, d\} \times \{l\}.$$

It is important to recognize that the term rationalization operator is not arbitrary; it is intended to reflect its function, capturing the behavioral implications of the players' rationality and their higher-order beliefs in rationality.

Assumptions on rationality and beliefs	Behavioral implications
R	$\rho(A) = br(\Delta(A))$
$R \cap \mathcal{B}(R)$	$\rho^2(A)$
$R \cap \mathcal{B}(R) \cap \mathcal{B}^2(R)$	$\rho^3(A)$
$\vdots$	$\vdots$
$R \cap C\mathcal{B}(R)$	$\rho^\infty(A)$

Figure 12 Epistemic assumptions and the rationalization operator

From the previous subsection, we saw that assuming rationality implies that players choose actions only from the set $\{u,d\}\times\{l,r\}$. Moreover, assuming that players believe that the other is rational further restricts choices to $\{u,d\}\times\{l\}$. By similar reasoning, this pattern continues for higher levels of belief in rationality.

In general, the sequence of sets $\rho^1(A), \rho^2(A), \ldots$ is monotonically decreasing, meaning

$$\rho^1(A) \supseteq \rho^2(A) \supseteq \ldots.$$

Thus, it is natural to define its limit as

$$\rho^\infty(A) = \bigcap_{i \in \mathbb{N}} \rho^i(A).$$

The action profiles in $\rho^\infty(A)$ are precisely what Pearce (1984) and Bernheim (1984) referred to as rationalizable.

Figure 12 summarizes the correspondence between different levels of mutual belief in rationality and successive iterations of the rationalization operator, where

$$\mathcal{B}^{n+1}E = \mathcal{B}\mathcal{B}^n E = \bigcap_{i \in N} \mathcal{B}_i \mathcal{B}^n E \quad \text{and} \quad C\mathcal{B}E = \mathcal{B}E \cap \mathcal{B}^2 E \cap \ldots$$

Having understood this correspondence, we should not be surprised by the next iteration of ρ in our game:

$$\rho^3(A) = \rho(\{u,d\} \times \{l\}) = \{u\} \times \{l\}.$$

One can also compute that we have reached a fixpoint of ρ in the game, that is, $\rho^{n+1}(A) = \rho^n(A)$ for $n \geq 3$. In fact, it can be proved that in any finite game there exists a positive integer k such that

$$\rho^{n+1}(A) = \rho^n(A) = \rho^\infty(A) \neq \emptyset \text{ for } n \geq k.$$

Both Bernheim (1984) and Pearce (1984) then proved the following result.

Theorem 4: *In any finite game, for every $n \geq 1$, $\rho^n(A) = NSD^n(A)$.*

We are now in position to state the fundamental theorem of epistemic game theory. An informal treatment can be found in Bernheim (1984) and Pearce (1984), and a formal presentation in Brandenburger and Dekel (1987) and Tan and da Costa Werlang (1988).

Theorem 5 (The Fundamental Theorem of Epistemic Game Theory): *An action profile* $a = (a_1, \ldots, a_n)$ *survives iterated elimination of strictly dominated actions if and only if there exists a type space (or, equivalently, a probabilistic Kripke model) in which the players choose the action profile* $a = (a_1, \ldots, a_n)$ *at states where rationality and common belief in rationality hold.*

4.4 Discussion

The fundamental theorem is fundamental for both historical and conceptual reasons. Historically, it marks the beginning of epistemic game theory as an alternative to the equilibrium refinement program. Pearce's contribution was originally framed within the refinement program.[29] His motivation, however, went deeper. He argued that rather than refining solution concepts based on intuitions about rational or irrational play, progress should come from a more principled foundation: the well-understood theory of rational decision-making under uncertainty.

This shift was quickly recognized as a research program in its own right (Werlang, 1986; Tan and da Costa Werlang, 1988), marking the formal emergence of epistemic game theory. Of course, as noted in Section 1, earlier contributions such as (Armbruster and Böge, 1979) and (Spohn, 1982) predate Pearce's work, with Spohn (1982) arguably making an even more explicit case for a methodological shift. Nonetheless, an examination of citation patterns in early epistemic game theory suggests that these earlier contributions were relatively less well-known.

The fundamental theorem is also fundamental for conceptual reasons. First and foremost, it pinpoints the behavioral consequences of rationality and common belief of rationality, which has been taken to be *the* fundamental assumptions of game theory, for example, (Myerson, 1991; Aumann and Dreze, 2008). The fundamental theorem shows, on the one hand, that rationality and common belief of rationality entails a well-known solution concept for static games, namely iterated elimination of strictly dominated actions. On

[29] See the quote from Pearce's abstract in Section 2.

the other hand, the fundamental theorem also shows that common belief of rationality does *not* correspond to Nash equilibrium. The fundamental assumptions of game theory fall short of explaining its central solution concept (c.f. Section 3 again).

Another conceptual reason why the fundamental theorem is indeed fundamental is that it provides the canonical structure of what has been called "epistemic conditions for" or "epistemic characterization of" solution concepts. An epistemic characterization result for a solution concept typically shows, on the one hand, that certain "epistemic conditions" are sufficient for rational players to play according to that solution concept. The fundamental theorem does exactly that: It shows that whenever there is common belief in rationality – this is the epistemic condition – among rational players, then they play a profile of actions that would survive iterated elimination of strictly dominated actions. On the other hand, epistemic characterization results show that any profile compatible with a given solution concept can be seen *as if* it was played under the given epistemic conditions. This is also the case for the fundamental theorem. Any profile of actions that survives iterated elimination of strictly dominated actions is rationalizable; that is, it can be seen as if it was played by rational players under common belief in rationality. The fundamental theorem is, in fact, not the first epistemic characterization result that we encounter. The characterizations of equilibrium play that we discussed in Section 3 are also epistemic characterizations in that sense.

The fundamental theorem shows the limited explanatory power of assuming rationality and common belief in rationality. These assumptions explain little in the sense that, for most games of interest, there are often few or no profiles that would be eliminated by iterated elimination of strictly dominated actions. Looking back at the different games that we introduced in Section 2, iterated elimination does not rule out any action in Battle of the Sexes (Figure 2) and Matching Pennies (Figure 3). The challenge in these games is, arguably, to select an equilibrium – or, in the case of Matching Pennies, to settle on the mixed-action equilibrium. However, assuming that players are rational and that this is common belief does not help resolve this challenge. The flipside of this observation is that rationality and common belief in rationality can also be seen as explaining too much. Any off-equilibrium profile that survives the elimination process is rationalizable. Moreover, even when iterated elimination of strictly dominated actions uniquely determines the outcome, the result may not always be convincing or intuitive – such as in the case of the Traveler's Dilemma (Basu, 1994) or the Prisoners' Dilemma (Kuhn, 2024).

4.5 Weak Dominance

While the fundamental theorem naturally bridges rationalizability and actions that would survive iterated elimination of strictly dominated actions, moving to *weakly* dominated actions complicates the picture. Recall that an action a_i of player i is weakly dominated if there is another action α_i such that, first, for any profile α_{-i} of choices of the others, the outcome of playing α_i against α_{-i} is at least as good as the outcome of playing a_i against α_{-i} and, second, there is *some* profile α_{-i} such that , playing α_i against α_{-i} is strictly better than the outcome of playing a_i against α_{-i}. An action that is not weakly dominated is often called "admissible." As it turns out, the combination of universal and existential quantification in the definition of admissibility – "for all" profiles . . . for "some" profiles – complicates the picture quite a lot.

To illustrate the difficulty, consider what happens in the game in Figure 13 when we try to iteratively eliminate weakly dominated actions. Action d is weakly dominated by u for Ida. After ruling out d, action r becomes strictly dominated by l for Joe, leading to the profile (u, l). Observe, however, that *d is not weakly dominated anymore* when we assume that Joe rules out r. If Joe plays l then Ida is indifferent between u and d. More generally, unlike strict dominance, admissibility is not *monotonic*: Some actions that are weakly dominated in a static game might not be weakly dominated when we move to a strict subgame.

For authors like Samuelson (1992), the non-monotonicity of admissibility has a paradoxical quality. One way to see this is by revisiting the relationship between dominance and best replies, as discussed in Section 2. We established that an action is strictly dominated if and only if it is never a best reply. Similarly, we saw a related result for weak dominance: An action is weakly dominated if and only if it is never a best reply to a *full-support* belief – that is, a belief that assigns nonzero probability to all possible action profiles of the other players. Such full-support beliefs can be considered "cautious" because, while they may assign very low probability to certain action profiles, they never categorically rule out any possibility.

The paradoxical nature of the non-monotonicity of weak dominance lies in the tension between the iterated elimination process and the type of cautious

	l	r
u	1, 1	1, 0
d	1, 0	0, 1

Figure 13 A two-player game from Samuelson (1992)

beliefs that underlie admissibility. Ruling out some actions on the ground that they are weakly dominated can naturally be seen as assigning probability zero to them. But this is precisely what is excluded by assuming that the players only play best replies to cautious beliefs.

From an epistemic point of view, this tension challenges the connection between iterated elimination of weakly dominated actions and what would be the choices of rational players with cautious beliefs and common belief of that fact. Some solutions to this challenge have been proposed, for example (Brandenburger et al., 2008; Halpern and Pass, 2009), but they are mathematically too involved to be presented in any detail in this Element. Intuitively, however, they require introducing a way for the players to assign probability zero to some actions of others, while not ruling them out completely. This is done by introducing sophisticated models of belief revision for static games, for example, the so-called *lexicographic probability systems* (see Halpern (2010)).

5 Correlations and Beliefs in Correlations

An important assumption in classical game theory is that players make decisions independently, based solely on their own information and strategy. In this section, we explore two ways to drop this assumption: Players might correlate their actions based on an external mediator, or they might believe that others are correlating their actions.

5.1 Correlated Equilibrium

The assumption that players make decisions independently can sometimes lead to suboptimal outcomes, as players are unable to coordinate their actions to achieve mutually beneficial results. To illustrate, recall the Battle of the Sexes game from Figure 2, where two players, Ida and Joe, must choose between two activities, going to a football match (f) or going the golf course (g). The game matrix is repeated as Figure 14 for convenience.

In each of the pure-action Nash equilibria, one player gets their preferred outcome while the other does not. For instance, if both choose to go to the football match (f, f), Ida gets 2 and Joe gets 1. Alternatively, if both go to the

Player 1 \ Player 2	f	g
f	2, 1	0, 0
g	0, 0	1, 2

Figure 14 The static game typically called the Battle of the Sexes

	f	g
f	$\frac{2}{3} \times \frac{1}{3} = \frac{2}{9}$	$\frac{2}{3} \times \frac{2}{3} = \frac{4}{9}$
g	$\frac{1}{3} \times \frac{1}{3} = \frac{1}{9}$	$\frac{1}{3} \times \frac{2}{3} = \frac{2}{9}$

Figure 15 Probability of each of the four outcomes

golf course (g, g), Ida gets 1 and Joe gets 2. However, if they cannot coordinate and end up choosing different activities, they both get a payoff of 0.

There is a mixed-action Nash equilibrium where Ida chooses f with probability $\frac{2}{3}$ and g with probability $\frac{1}{3}$, while Joe plays f with probability $\frac{1}{3}$ and g with probability $\frac{2}{3}$. Imagine that Ida and Joe each use a randomizing device to select their actions based on this mixed-action Nash equilibrium. Ida flips a biased coin that lands Heads (H) with probability $\frac{2}{3}$ and then chooses f if the coin lands H and g otherwise. Similarly, Joe flips a biased coin that lands Tails (T) with probability $\frac{2}{3}$ and then chooses g if the coin lands T and f otherwise. Since the two coin flips are independent, the probability of each of the four outcomes is given in Figure 15.

Given the previous probability table, the expected payoffs for the players are:

- Ida's expected payoff is $\frac{2}{9} \times 2 + \frac{4}{9} \times 0 + \frac{1}{9} \times 0 + \frac{2}{9} \times 1 = \frac{2}{3}$.
- Joe's expected payoff is $\frac{2}{9} \times 1 + \frac{4}{9} \times 0 + \frac{1}{9} \times 0 + \frac{2}{9} \times 2 = \frac{2}{3}$.

Note that the expected payoffs for both players are less than 1, which is the payoff the players would get in their least preferred Nash equilibrium.

Now, suppose that instead of each player flipping a coin independently, a "trusted source" flips a fair coin and, based on the outcome of the coin toss, tells the players what they should do:

- If the coin lands Heads, it is recommended that Joe and Ida both play f.
- If the coin lands Tails, it is recommended that Joe and Ida both play g.

The players are under no obligation to follow the recommendation. However, if both players know the aforementioned protocol and they are rational, they will follow the recommendation. The crucial observation is that if Ida receives the recommendation to play f, she knows that Joe received the same recommendation, and so deviating from the recommendation will result in a payoff of 0 (assuming that Joe follows the recommendation). Thus, the probability distribution over the outcomes changes (Figure 16).

In this case, the expected payoffs for both players increase:

- Ida's expected payoff is $\frac{1}{2} \times 2 + 0 \times 0 + 0 \times 0 + \frac{1}{2} \times 1 = \frac{3}{2}$.
- Joe's expected payoff is $\frac{1}{2} \times 1 + 0 \times 0 + 0 \times 0 + \frac{1}{2} \times 2 = \frac{3}{2}$.

	f	g	
f	$\frac{1}{2}$	0	w
g	0	$\frac{1}{2}$	

Figure 16 The new probability distribution over the outcomes

This example highlights that by using a correlation device, players can achieve higher expected payoffs. The concept of a *correlated equilibrium* captures this idea by allowing for such coordination. Intuitively, a correlated strategy can be viewed as a mediator that selects an action profile a according to some probability measure $\mu \in \Delta(A)$ and then tells each player $i \in N$ their action in a. A correlated strategy μ is a correlated equilibrium when the action revealed to each player i is a best reply to the belief that i has about the other actions, given the message received from the mediator and μ. In the example described earlier, μ is the probability that assigns 0.5 to (f, f) and 0.5 to (g, g).

Aumann (1987) characterized correlated equilibrium using a variant of probabilistic Kripke models.[30] Given a static game G, a *prior probabilistic Kripke model* for G is a tuple $(W, (p_i)_{i\in N}, \mathrm{s})$, where W and $\mathrm{s} : W \to A$ are as in probabilistic Kripke models for G from Section 2, while p_i is a probability measure on W for each $i \in N$, representing player i's prior belief. It is assumed that for all action profiles $a \in A$, $[a]$ is measurable according to p_i for each $i \in N$, and for all $a_i \in A_i$ and $i \in N$, $p_i([a_i]) > 0$. There are two key observations needed to understand Aumann's result:

- Every prior probabilistic Kripke model for a game G generates a probabilistic Kripke model for G. Suppose that $(W, (p_i)_{i\in N}, \mathrm{s})$ is a prior probabilistic Kripke model for some game G. We generate a probabilistic Kripke model for G, $(W, (b_i)_{i\in N}, \mathrm{s})$, so that the model has the same set W and function s, where b_i is defined as follows: For all $w \in W$,

  $$b_i(w)(\cdot) = p_i(\cdot \mid [\mathrm{s}_i(w)]).$$

  So, i's posterior belief at a state w is her prior p_i conditioned on the event that she chooses the action played at w.
- Every prior probabilistic Kripke model for a game G generates a probability on the set of action profiles for each player $i \in N$. Consider a static game G and a prior probabilistic Kripke model $(W, (p_i)_{i\in N}, \mathrm{s})$ for G. For all $i \in N$, let $\beta_i^p \in \Delta(A)$ be defined as $\beta_i^p(a) = p_i([a])$ for all $a \in A$.

The following result by Aumann provides an illuminating perspective on correlated equilibrium.

[30] Our presentation of this result is based on Halpern and Moses (2017).

Theorem 6 (Aumann (1987)): *Consider a static game G. Then, $\mu \in \Delta(A)$ is a correlated equilibrium if, and only if, there exists a prior probabilistic Kripke model $(W, (p_i)_{i \in N}, \mathrm{s})$ for G such that for all $i \in N$, $\beta_i^p = \mu$, and in the generated probabilistic Kripke model $(W, (b_i)_{i \in N}, \mathrm{s})$, $R_i = W$ for all $i \in N$ (i.e., the players are rational at all states).*

Note that the prior probabilistic Kripke model mentioned in the previous theorem has the condition that for all $i \in N$, $\beta_i^p = \mu$. Thus, all players have the same prior beliefs about the profiles in the game. We say that a prior probabilistic Kripke model $(W, (p_i)_{i \in N}, \mathrm{s})$ has a *common prior* when, for all $i, j \in N$, $p_i = p_j$. Thus, Aumann characterizes correlated equilibrium in models with common prior. Since the players are rational in every state of the model, it follows that there is common knowledge of their rationality. Moreover, if the common prior μ can be expressed as a product measure $\mu(a) = \Pi_{i \in N} \mu_i(a_i)$, where $\mu_i \in \Delta(A_i)$, then μ is, in fact, a Nash equilibrium. Therefore, as we briefly mentioned in Section 3, the aforementioned theorem also provides an alternative route to understanding the epistemic interpretation of mixed actions.

5.2 Intrinsic Correlation

In the previous section, we discussed correlation that relies on an outside mediator. However, when there are three or more players, another form of correlation becomes important: Players may *believe* that the choices of others are correlated. Such beliefs can be crucial for rationalizing outcomes in certain games (Brandenburger and Dekel, 1987; Brandenburger and Friedenberg, 2008). Consider, for example, the three-person game depicted in Figure 17 from Brandenburger and Friedenberg (2008), where Ida chooses the rows (u or d), Joe the columns (l or r), and Kai the matrices (x, y, or z).

Observe that no action in the game in Figure 17 is strictly dominated.[31] In particular, for Kai, action y is preferable to z if Ida and Joe choose u and l, respectively, and preferable to x if Ida and Joe choose d and r, respectively, and there exists a belief $\beta_k \in \Delta(A_i \times A_j)$ of Kai such that y is a best response

x:

	l	r
u	1, 1, 3	1, 0, 3
d	0, 1, 1	0, 0, 0

y:

	l	r
u	1, 1, 2	1, 0, 0
d	0, 1, 0	1, 1, 2

z:

	l	r
u	1, 1, 0	1, 0, 0
d	0, 1, 3	0, 0, 3

Figure 17 A three-player game from Brandenburger and Friedenberg (2008)

31 Of course, both d and r are weakly dominated by u and l, respectively.

to β_k. The reader can verify that $\beta_k(u,l) = \beta_k(d,r) = 0.5$ is such a belief. With this belief, however, Kai thinks that Ida and Joe's actions are correlated, expecting them to coordinate on either (u,l) or (d,r).

A bit of algebra reveals that there is no belief of Kai under which y is a best reply if Kai's belief can be decomposed into two independent beliefs about Ida's and Joe's choices, respectively (Brandenburger and Friedenberg, 2008). In other words, Kai's choice of y can only be rationalized if he believes that Ida's and Joe's choices are correlated.

As noted in the introduction to this section, players choose their actions independently. Since there is no external mechanism to correlate the players' choices, how did Kai arrive at the belief that Ida and Joe's actions are correlated? Brandenburger and Friedenberg (2008) suggest an answer:

> Under the epistemic approach, a complete description of a game includes specifying not only the players' strategy sets and payoff functions, but also their hierarchies of beliefs (beliefs, beliefs about beliefs, etc.) about the strategies played in the game. This gives a source of correlation. A player can think that other players' strategy choices are correlated, because he thinks what they believe about the game is correlated. (p. 29)

The idea is that Kai believes Ida and Joe's actions are correlated because Ida and Joe share the same beliefs about the game. To make this more formal, consider a type space for the game G in Figure 17 with a type t_1 for Ida, t_2 for Joe, and t_3 for Kai (with b_1, b_2, and b_3 being the belief maps for the players). Suppose that Kai has the following beliefs about Ida and Joe's actions: $b_3(t_3)(u,l) = \frac{1}{2}$ and $b_3(t_3)(d,r) = \frac{1}{2}$. This means that Kai believes Ida and Joe's actions are correlated. If this correlation does not arise from an external mediator, then the following two constraints should be satisfied:

- Kai's belief that Ida and Joe's actions are correlated arises from uncertainty about Ida and Joe's hierarchy of beliefs. If this uncertainty is resolved, then, since there is no external mediator sending signals to the players, Kai should no longer believe that Ida and Joe's actions are correlated. This can be expressed as:

  $$b_3(t_3)(u,l \mid t_1,t_2) = b_3(t_3)(l \mid t_1,t_2) \times b_3(t_3)(l \mid t_1,t_2),$$

  where, for instance, $b_3(t_3)(u,l \mid t_1,t_2)$ is the probability that Kai assigns to Ida playing u and Joe playing l given that Ida is type t_1 and Joe is type t_2. This leads to the following constraint on type spaces:

  Conditional Independence (CI): The measure associated with each type assesses the other players' action choices as independent, conditional on their hierarchies of beliefs.

- Kai's belief about Ida's action choice depends only on Ida's type. Thus, learning Joe's type in addition to Ida's type should not change the probability that Kai assigns to Ida playing u after learning Ida's type:

$$b_3(t_3)(u \mid t_1, t_2) = b_3(t_3)(u \mid t_1) \quad b_3(t_3)(d \mid t_1, t_2) = b_3(t_3)(d \mid t_1),$$

  and *mutatis mutandis* for Joe. This suggests the following constraint on type spaces:
  Sufficiency (SUFF): Learning the type of a player is sufficient for the other player to fix the probabilities of the actions for that player.

Recall from Section 4 that in any finite static game G, the rationalizable actions are exactly the actions that are played at any state in a game model where the players are rational and there is common belief of rationality. The game in Figure 17 shows that assuming rationality and common belief of rationality sometimes requires a player to believe that the actions of the other players are correlated. This raises an interesting question about the relationship between rationality, common belief in rationality, and the constraints of Conditional Independence and Sufficiency. Do the assumptions of CI and/or SUFF further restrict the actions played, assuming that the players are rational and there is a common belief in rationality? Brandenburger and Friedenberg (2008) showed that the answer to this question is yes. This result can be viewed as an impossibility theorem. It shows that it is impossible to find a type space of a game satisfying rationality and common belief of rationality and either of the constraints CI or SUFF.

Theorem 7 (Brandenburger and Friedenberg (2008)): *There is a game G and a rationalizable action profile a in G such that no state in a type space for G satisfies rationality and common belief of rationality and CI. There is an analogous result for SUFF.*

The proof of this theorem is beyond the scope of this Element (see Brandenburger and Friedenberg (2008) for details). We conclude with a brief discussion on the importance of this theorem.

The fundamental theorem of epistemic game theory (discussed in Section 4) characterizes the actions that are consistent with rationality and common belief of rationality. When there are more than three players, the previous example shows that this may require that the players believe that their opponents' actions are correlated. Believing that some actions are correlated requires a "common-cause explanation." In the previous section, the common cause of correlated actions is an outside mediator sending signals to the players. The epistemic perspective on games presented in this Element suggests another common-cause explanation: Players believe that the other players are thinking about the game in the same way.

From the epistemic perspective, CI and SUFF are natural assumptions. These conditions together imply that if a player believes that the types of the other players are independent, then the player must also believe that the actions of the other players are independent. In other words, the only reason to believe that a group of players' actions are correlated is that they are all thinking about the game in the same way. Theorem 7 is an important result that highlights the limits of this assumption. It shows that there are some games for which it is impossible to simultaneously satisfy rationality, common belief of rationality, and the constraint that the only source of correlation is the players' beliefs about what other players are going to do in the game.

6 Dynamic Games and Subgame Perfect Equilibrium

In this section, we discuss dynamic games, in which players make decisions sequentially rather than simultaneously. Many familiar board games, such as chess or checkers, fall into this category, as do strategic interactions like bargaining, trading, and auctions. The most well-known solution concept for dynamic games is the *subgame perfect equilibrium* (Selten, 1975), which is typically computed using the *backward induction algorithm*. The epistemic foundation of subgame perfect equilibrium has been a long-standing topic of debate in epistemic game theory. The central question concerns whether common knowledge of rationality necessarily implies that players will choose according to subgame perfect equilibrium. Toward the end of this section, we will also outline epistemic approaches to an alternative class of solution concepts for dynamic games, those based on forward induction reasoning.

6.1 Dynamic Games, Subgame Perfect Equilibrium, Backward Induction

The key distinction between dynamic and static games is that dynamic games, also known as extensive-form games, explicitly describe the order in which players move. We focus on dynamic games with *perfect information* – games in which there are no simultaneous choices and no uncertainty about past moves. Examples include chess and checkers, as well as many trust games studied in philosophy, such as those in Gauthier (1987). By contrast, games of imperfect information include cases where players lack knowledge of past moves. For example, in most card games, such as poker or bridge, players know their own hands but not those of their opponents.[32]

A *dynamic game with perfect information* is a tuple $(N, T, Act, \tau, (\pi_i)_{i \in N})$, where

- N is a finite set of players;
- T is a tree describing the order of choices for each player consisting of a set of nodes and an immediate successor relation $\rightarrowtail$ between nodes.
  - When a node v' immediately follows a node v, we write $v \rightarrowtail v'$ and call v' a *successor* of v.
  - Let Z be the set of nodes without any successors (called *terminal nodes*) and V the remaining nodes (called *decision nodes*). We write v_0 for the *root* of the tree, which is a node without predecessors. By assumption, there is only one root.
  - Each transition from a node v to a successor node v' is labeled by an action. We write $A(v)$ to denote the set of actions available at v.
- τ is a turn function that assigns a player to each decision node $v \in V$. For each player $i \in N$, we write $V_i = \{v \in V \mid \tau(v) = i\}$ for the set of nodes where it is i's turn to act.
- $\pi_i : Z \to \mathbb{R}$ is the payoff function for player i assigning real numbers to each terminal node.

In dynamic games, it is important to distinguish between actions and *strategies*. While "strategy" commonly means a broad long-term plan, in game theory it refers to a complete plan specifying a player's chosen action at every possible decision point – even those that would never be reached when following the plan itself. Formally, a strategy is defined as follows. Suppose that G is a dynamic game. A strategy s_i for player i in G is a function s_i that maps each $v \in V_i$ to an element of $A(v)$. For each player i, we use S_i to denote the set of all possible strategies of i. A strategy profile for G, denoted s, is an element of $\times_{i \in N} S_i$. Given a strategy profile s, we write s_i for player i's component of s and s_{-i} for the sequence of strategies for all players except i.

Each strategy profile s for a dynamic game G generates a sequence of nodes $v_0, v_1, \ldots, v_k$, called a *path*, where v_k is a terminal node, and for all $0 \leq l < k$, $v_{l+1} = s_{\tau(v_l)}(v_l)$. We say that v is reached by a strategy profile s if v is on the

32 The key distinction between perfect and imperfect information games is that in the latter, uncertainty about past or present moves is inherent to the game's structure. This uncertainty does not arise from players forgetting or misrepresenting past actions; it would persist even if players had fully idealized cognitive capacities.

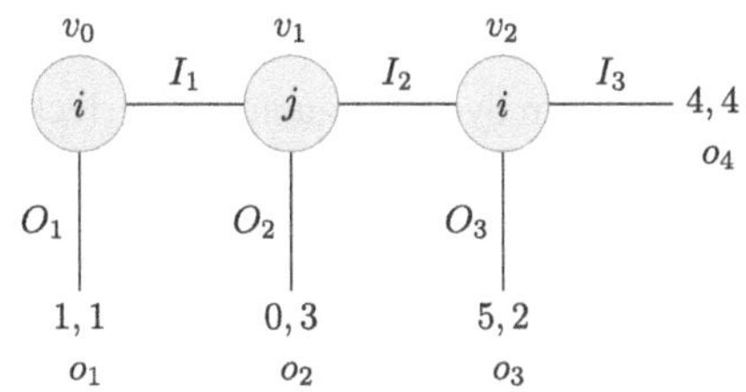

Figure 18 A Centipede game

path generated by s. Suppose that v is any node in a dynamic game. We write $out(v, s)$ for the terminal node that is reached if, starting at node v, all players move according to their respective strategies in the profile s. Given a decision node $v \in V_i$ for player i, a strategy $s_i \in S_i$ for player i, and a set $X \subseteq S_{-i}$ of strategy profiles of the opponents of i, let $Out_i(v, s, X)$ be the set of terminal nodes that may be reached if, starting at node v, player i uses strategy s_i and i's opponents use a strategy profile from X, that is, $Out_i(v, s, X) = \{out(v, (s, s_{-i})) : s_{-i} \in X\}$.

Figure 18 shows an example of a dynamic game, called the *centipede game* (Rosenthal, 1981). The decision nodes for Ida (i) and Joe (j) are $V_i = \{v_0, v_2\}$ and $V_j = \{v_1\}$. The terminal nodes are $Z = \{o_1, o_2, o_3, o_4\}$. The labels of the edges are the actions of each player. For example, $A(v_0) = \{O_1, I_1\}$. There are four strategies for Ida and two strategies for Joe. To simplify notation, we denote the players' strategies by the sequence of choices at each of their decision nodes. For example, Ida's strategy s_i^1 defined as $s_i^1(v_0) = O_1$ and $s_i^1(v_2) = O_3$ is denoted by the sequence O_1O_3. Thus, Ida's strategies are $s_i^1 = O_1O_3$, $s_i^2 = O_1I_3$, $s_i^3 = I_1O_3$ and $s_i^4 = I_1I_3$. Note that Ida's strategy s_i^2 specifies a move at v_2, although the earlier move at v_0, O_1, means that Ida will not have the chance to move at v_2. Similarly, Joe's strategies are $s_j^1 = O_2$ and $s_j^2 = I_2$. Then, for example, $out(v_1, (s_i^2, s_j^2)) = o_4$. Finally, if $X = \{s_i^1, s_i^4\}$, then $Out_j(v_1, s_j^2, X) = \{o_3, o_4\}$.

We are now ready to define subgame perfect equilibrium. Given a dynamic game G and a decision node v, the subgame generated by v is the dynamic game defined by restricting G to v and all nodes reachable from v (so v is the root in the subgame of G generated by v). For instance, in addition of the full game in Figure 18, there are two subgames: one generated by v_2 and the other generated by v_3. A strategy profile s is a subgame perfect equilibrium of G if, for any subgame of G, no player i has an incentive to deviate from s_i, given that the other players are following s_{-i}. That is, for any decision node v there is no $s_i' \in S_i$ such that $\pi_i(out(v, (s_i', s_{-i}))) > \pi_i(out(v, s))$ for all $i \in N$. For the game in Figure 18 the profile (O_1O_3, O_2) is the unique subgame perfect equilibrium.

The so-called backward induction algorithm can be used to compute the subgame perfect equilibrium in dynamic games of perfect information:[33]

Backward Induction Algorithm: Initially, all terminal nodes are marked with the players' payoffs. At a nonterminal node v, once all immediate successors are marked, the node is marked as follows: Find the immediate successor v' that has the highest payoff for the player whose turn it is to move at v, that is, $\tau(v)$. Copy the payoffs from v to v'. Repeat this procedure until all nodes are marked with the players' payoffs.

Given a dynamic game where all nodes are marked, the unique path that leads from the root v_0 to the outcome with the utilities that match the utilities assigned to v_0 is called the backward induction path. The backward induction algorithm defines, for every nonterminal node, a path from that node to a terminal node. These paths can be used to define strategies for each player: At each decision node v, choose the action that is consistent with the path from v. The resulting combination of strategies is the *backward induction profile*. This profile is a subgame perfect equilibrium.

We illustrate the backward induction algorithm on the game in Figure 18:

- Node v_2 only has marked nodes as successors. Ida plays at node v_2, so we mark v_2 with $(5,2)$ since the outcome o_3 has the greatest payoff for Ida among all marked immediate successor nodes.
- Node v_1 only has marked nodes as successors. Joe plays at node v_1, so we mark v_1 with $(0,3)$ since the outcome o_2 has the greatest payoff for Joe among all marked immediate successor nodes.
- Node v_0 only has marked nodes as successors. Ida plays at node v_0, so we mark v_0 with $(1,1)$ since the outcome o_1 has the greatest payoff for Ida among all marked immediate successor nodes.

We can extract the strategy profile (s_i^1, s_j^1) from this labeling, which is indeed the subgame perfect equilibrium. Note that the subgame perfect equilibrium *path* from v_0 is O_1, since Ida ends the game at the first move.

[33] In this Element, we focus on dynamic games with perfect information where there are no "relevant ties" (Battigalli, 1997) – that is, no terminal nodes that yield the same payoff for player i when the last common predecessor node is one where i is to move. This assumption significantly simplifies the analysis, as it ensures that the game always has a unique subgame perfect equilibrium. Moreover, the key epistemic issues underlying this solution concept already emerge in this simple case. That said, backward induction reasoning extends beyond this restricted setting. It applies more broadly to dynamic games with imperfect or incomplete information as well, as discussed in Bonanno (2014); Perea (2014a); Catonini and Penta (2022).

6.2 Models of Dynamic Games

Suppose that $G = (N, T, Act, \tau, (\pi_i)_{i\in N})$ is a dynamic game. An *epistemic model of* G is a tuple

$$(W, (\Pi_i)_{i\in N}, \sigma)$$

where

- W is a nonempty set of states of the world;
- For each $i \in N$, Π_i is a partition of W;
  - We write $\Pi_i(w)$ for the element of Π_i that contains w.
- $\mathrm{s} : W \to \times_{i\in N} S_i$ is a function that assigns to each state w, a strategy profile for G. If $\mathrm{s}(w) = s$, then we write $\mathrm{s}_i(w)$ for s_i and $\mathrm{s}_{-i}(w)$ for s_{-i}.

As usual, we assume that the players know their own strategies. This boils down to the assumption that for all $w \in W$, if $w' \in \Pi_i(w)$, then $\mathrm{s}_i(w) = \mathrm{s}_i(w')$.

Suppose that G is a dynamic game with perfect information and $(W, (\Pi_i)_{i\in N}, \mathrm{s})$ is an epistemic model of G. We can now introduce the following key notions needed to define common knowledge of rationality.

- For an event $E \subseteq W$, the event that *player* i *knows that* E is defined as $\mathcal{K}_i(E) = \{w : \Pi_i(w) \subseteq E\}$.
- Common knowledge of an event $E \subseteq W$, denoted $\mathcal{CK}(E)$, is defined as in Section 2:

  $$\mathcal{CK}(E) = \bigcap_{n\geq 1} \mathcal{K}^n(E)$$

  with $\mathcal{K}^{n+1}(E) = \mathcal{K}(\mathcal{K}^n(E))$ and $\mathcal{K}(E) = \bigcap_{i\in N} \mathcal{K}_i(E)$.
- For a state w, let $S_{-i}(w) = \{\mathrm{s}_{-i}(w') : w' \in \Pi_i(w)\}$ be the set of strategy profiles of player i's opponents that i thinks are possible at w. Then, $Out_i(v, s, S_{-i}(w))$ is the set of terminal nodes that player i considers reachable starting from node v if she follows strategy s. Player i is *rational at node* $v \in V_i$ *in state* w provided that, for all strategies $s \in S_i$ such that $s \neq \mathrm{s}_i(w)$, there are terminal nodes $o' \in Out_i(v, s, S_{-i}(w))$ and $o \in Out_i(v, \mathrm{s}_i(w), S_{-i}(w))$ such that $\pi_i(o) \geq \pi_i(o')$. Thus, player i is rational at a decision node $v \in V_i$ in state w provided that i does not know that there is an alternative strategy that would always give her a strictly higher payoff.[34]

[34] In the previous sections we understood rationality as playing a best reply in the sense of maximization of expected payoffs. Since there is no probabilistic beliefs in the models we use in this section, we follow the literature in epistemic game theory and use the present definition of rationality at a node. Note that this notion is weaker than playing a best reply given a probabilistic belief. An action that is not rational in the present sense is never a best reply, but there

- Player i is *substantively rational at state* w when for all decision nodes $v \in V_i$, i is rational at v in state w. Note that player i is substantially rational in state w when i is rational at *all* of i's decision nodes, even those that are ruled out at previous decision nodes of player i according to i's strategy at w.[35]

The event that player i is substantively rational is defined as follows

$$SR_i = \{w : \text{player } i \text{ is substantively rational at state } w\},$$

and the event that all players are substantively rational is $SR = \bigcap_{i \in N} SR_i$. In the remainder of this section, "common knowledge of rationality" means common knowledge of substantive rationality (i.e., $\mathcal{CK}(SR)$).

6.3 Common Knowledge of Rationality and Subgame Perfect Equilibrium

One of the oldest questions in the epistemic analysis of dynamic games is whether common knowledge of rationality is sufficient for subgame perfect equilibrium. The literature presents seemingly contradictory mathematical results, with some proving sufficiency and others disproving it. These divergent conclusions stem from different assumptions about how the players change their minds when they observe something unexpected.[36]

6.3.1 Common Knowledge of Rationality Is NOT Sufficient for Subgame Perfect Equilibrium

Arguments against the sufficiency of common knowledge of rationality for subgame perfect equilibrium in perfect information games fall into two categories:

1. The first group argues that common knowledge of rationality is *incoherent* at nodes outside the subgame perfect equilibrium path. This position was developed by Bicchieri (1988a), Basu (1990), and Reny (1988, 1993).
2. The second group argues that common knowledge of rationality is coherent even at nonequilibrium nodes, but it is not sufficient to guarantee subgame perfect equilibrium. That is, they claim that rational players may adopt

could be actions that are not best replies to a given probabilistic belief, but are rational in the present sense.

35 This, of course, raises the question of what the players should believe about their own and the others' rationality and choices at those counterfactual nodes. We come back to this point extensively in the next section.

36 The presentation in this section closely follows Kuechle (2009), which we strongly recommend for further details.

strategies outside subgame perfect equilibrium despite there being common knowledge of rationality. This view originated with Binmore (1987) and Bicchieri (1988b), and has been formalized in subsequent literature by Bonanno (1991); Aumann (1998); Stalnaker (1996, 1999).

We illustrate the arguments in the first group using the centipede game in Figure 18. Observe first that if both players are rational, Joe knows that Ida is rational, Ida knows that Joe is rational, and Ida knows that Joe knows that she is rational, then Ida will choose O_1 at her first decision node v_0. Why? No matter what she believes or knows, if Ida is rational then she would never play I_3 at her final decision node v_2. Since Joe knows that Ida is rational and her only rational choice at v_2 is O_3, Joe knows that Ida will choose O_3 at v_2. Thus, if he is rational, then he will play O_2 at his decision node v_1. But then, if Ida knows that Joe is rational, and she also knows that Joe knows that she is rational, then Ida knows that Joe will choose his only rational choice O_2 at v_1. Thus, since Ida is rational, Ida will choose O_1 at v_0.

Bicchieri (1988a) and Reny (1988, 1993) have argued, however, that this reasoning breaks down if we assume that rationality is common knowledge. Suppose that, alongside the assumptions we made in the previous paragraph, Joe knows that Ida knows that he is rational, and he knows that Ida knows that he knows that she is rational. Then Joe also knows that the only rational choice for Ida at the first node is O_1. Since he knows that Ida is rational, he also knows that the decision node v_1, where he was planning to play O_2, will *not* be reached. That is, if Joe knows that v_1 has been reached, then he must conclude that Ida's choice at the previous node was not compatible with her being rational, or her knowing that he is rational, or her knowing that he knows that she is rational. So Joe cannot simultaneously know that rationality is common knowledge and that v_1 has been reached. Hence, common knowledge of rationality is incoherent at such off-equilibrium nodes, such as v_1.

This argument boils down to the observation that Joe knowing what Ida knows at the beginning of the game appears to undermine the very reasoning that leads to the subgame perfect equilibrium. Reny (1992) strengthened this observation by arguing that common knowledge of rationality is consistent only in games where *every* node is reached by the subgame perfect equilibrium.[37]

These impossibility results can be reinterpreted in a way that supports the second group's argument: Common knowledge of rationality may be consistent with the players choosing a subgame perfect equilibrium, but is not sufficient to

[37] Basu (1990) and later de Bruin (2010) provide similar results.

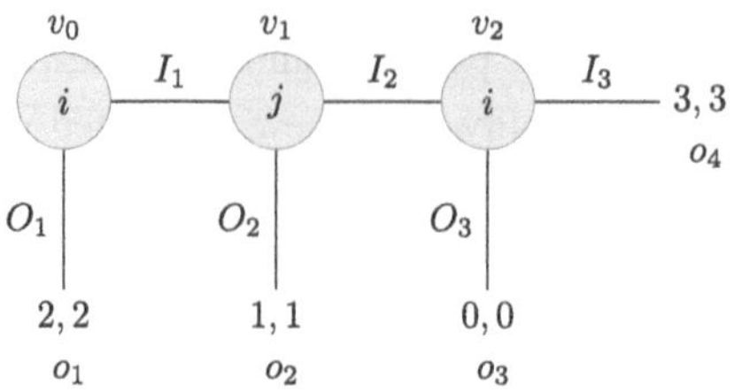

Figure 19 A dynamic game

ensure these choices. The key insight is that observing deviations from subgame perfect equilibrium strategies may trigger revisions in players' expectations about others' future actions. These revised expectations may then rationalize departures from the subgame perfect equilibrium profile.

To explain this argument consider the dynamic game in Figure 19, developed by Halpern (2001) in order to illustrate the result by Stalnaker (1999). The subgame perfect equilibrium profile is $(I_1 I_3, I_2)$ leading to the outcome o_4 with both players receiving a payoff of 3. Now consider the following epistemic model for this game. The set of states is $W = \{w_1, w_2, w_3, w_4, w_5\}$ with s defined as follows:

- $\mathrm{s}(w_1) = (O_1 I_3, O_2)$
- $\mathrm{s}(w_2) = (I_1 I_3, O_2)$
- $\mathrm{s}(w_3) = (I_1 O_3, O_2)$
- $\mathrm{s}(w_4) = (I_1 I_3, I_2)$
- $\mathrm{s}(w_5) = (I_1 O_3, I_2)$

We assume that, in this model, Ida always knows what state she is in: $\Pi_i(w_k) = \{w_k\}$ for all $k \in \{1, 2, 3, 4, 5\}$. In states w_1, w_4, and w_5, Joe knows what state he is in ($\Pi_j(w_k) = \{w_k\}$ for $k \in \{1, 4, 5\}$), but he cannot distinguish states w_2 and w_3 ($\Pi_j(w_2) = \Pi_j(w_3) = \{w_2, w_3\}$).

We will argue that the players are rational at state w_1, and so since $\Pi_i(w_1) = \Pi_j(w_2) = \{w_1\}$, there is common knowledge of rationality. However, since $\mathrm{s}(w_1) = (O_1 I_3, O_2)$, the players are *not* playing the subgame perfect equilibrium. The key to this argument is to allow Joe's knowledge to change when considering what he will believe if v_1 is reached in state w_1.

Stalnaker (1996, 1999) and Halpern (2001) use *selection functions* to represent how players change their minds upon observing deviations from a fixed strategy profile.[38] A selection function for an epistemic model $(W, (\Pi_i)_{i \in N}, \mathrm{s})$ is a function $f : W \times V \to W$, mapping pairs (w, v) consisting of a state

[38] See also Bicchieri (1988b) and (Samet, 1996) for a similar approach to representing how the players revise their beliefs in a dynamic game.

$w \in W$ and a decision node v from T to a state $f(w,v) \in W$. Intuitively, $f(w,v) = w'$ means that if the players suppose that node v is reached in state w, then the players' information would change from what is described at state w to what is described at w'. Of course, not every selection function captures *rational* revision policies. Stalnaker (1996) imposes the following three postulates for the selection function f, inspired by classical belief revision theory (Alchourrón et al., 1985):

1. *Success*: The node v is reached[39] in $f(w,v)$. Revision is successful in the sense that what is assumed to be true, namely that v is reached, is indeed the case at the new information state.
2. *Centering*: If v is reached in w then $f(w,v) = w$. The players do not change their mind if the incoming information is already consistent with what they know at the current state.
3. *Minimality*: $\mathrm{s}(f(w,v)) = \mathrm{s}(w)$ for the subtree starting at v. The players make the minimal changes that are consistent with the new information, and in particular keep their hypotheses about what the others will play from node v onward.

The key step is to adapt the definition of substantive rationality to take into account the possibility that players may change what they know about the other players in response to observing their choices at previous nodes. Player i is *substantively rational* in state* w provided for all decision nodes $v \in V_i$, i is rational at v in state $f(w,v)$.

Returning to the epistemic model described earlier for the game in Figure 19, the reader can check that there is a unique selection function f for this model satisfying the three postulates earlier, such that $f(w_1,v_1) = w_2$ and $f(w_1,v_2) = w_4$. This means that, from the perspective of w_1, if v_2 was reached then Ida would still play I_3, and this would be common knowledge since $\Pi_i(w_4) = \Pi_j(w_4) = \{w_4\}$. However, if v_1 was reached in w_1, then Joe would still play O_2. Observe that this choice is *not* irrational for Joe. The reason is that $f(w_1,v_1) = w_2$ and, at w_2, we have that $\Pi_j(w_2) = \{w_2,w_3\}$, and so he considers it possible that Ida will play either I_3 or O_3. Thus, it is not irrational for Joe to play O_2. Thus, both Ida and Joe are substantively rational* at w_1 (and this is common knowledge).

The key to the previous argument lies in modeling belief revisions triggered by deviations from subgame perfect equilibrium. This argument does not rule

[39] A node v is *reached in a state* w provided that v is on the path generated by the strategy profile $\mathrm{s}(w)$ starting at the root note v_0.

out, however, that the subgame perfect equilibrium *path* is followed, even without players adopting the complete equilibrium strategy *profile*. In those cases the outcome of the game would be the same. Differences would arise in counterfactual scenarios. Indeed, Bonanno (1991) and Aumann (1998) have shown that common knowledge of rationality entails the backward induction path will be played. Moreover, Rich (2015) has shown that by strengthening the rationality constraints on the players' belief revision policies, players will select the full subgame perfect equilibrium profile.

6.3.2 Common Knowledge of Rationality IS Sufficient for Subgame Perfect Equilibrium

The limitations of common knowledge of (substantive) rationality discussed in the previous section can be overcome by using common knowledge of *current* and *future* rationality instead of common knowledge of rationality at *all* nodes. Rationality at a node v in state w, knowledge, and common knowledge are defined as in Section 6.2. Common knowledge of current and future rationality at node v in state w means that for any node v' in the subtree generated at v, it is common knowledge that all players are rational at v' in w.

This definition differs from substantive rationality, which requires players to be rational at *all* nodes in the game tree. Current and future rationality is a weaker notion, requiring only that players be rational from the current point forward. Of course, common knowledge of substantive rationality implies current and future common knowledge of rationality at all nodes, but the converse is not true.

The idea that common knowledge of future rationality entails that the players choose their components of the subgame perfect equilibrium has been formalized in different frameworks.[40] In fact, within epistemic game theory, common knowledge of future rationality has become the standard explanation for subgame perfect equilibrium.[41]

---

[40] See Perea (2007a) and Kuechle (2009) for detailed overviews of these arguments.

[41] One might have reservations using knowledge instead of the weaker notion of belief to characterize subgame perfect equilibrium. Knowledge, after all, might be seen as too demanding. There are two points to be made here. First, using knowledge instead of belief actually strengthens the arguments presented in the previous section, to the effect that even common knowledge of rationality is not sufficient for subgame perfect equilibrium. Second, to obtain sufficient conditions, one could replace the knowledge requirement with the conjunction of rationality and common belief in future rationality. However, as we saw earlier, knowledge is understood as correct certainty. There is no difference here between assuming rationality and common belief of future rationality, on the one hand, and assuming common correct certainty about rationality, on the other.

We illustrate the implications of assuming common knowledge of current and future rationality with the game in Figure 18. We already argued that, at v_2, Ida will choose O_3, assuming that she is rational. Now, at node v_1, if Joe is rational and knows that Ida *will* (in the future) be rational and thus choose O_3, then he will choose O_2. Finally if, at v_0, Ida knows that Joe will be rational at v_1, and that he knows that she will be rational at v_2, then if she is rational at v_0 then she will indeed choose O_1.

The crucial point is that Joe decides what to do at v_1 on the basis of what he expects Ida to do at v_2. The arguments discussed in Section 6.3.1 pointed out that if Joe starts wondering how v_1 could have been reached, he might be forced to reconsider whether Ida was rational or knew that he is rational. Focusing on current and future rationality avoids this altogether. At v_1, Joe does not take into account how that node has been reached; he just assumes that from now on Ida will be rational.

A similar point can be made using the game in Figure 19. Recall that in the epistemic model of this game, at state w_1 Joe is playing O_2, which is not his subgame perfect equilibrium strategy. Common knowledge of future rationality should fail at w_1. At v_2, the only rational choice for Ida is I_3. This is independent of what she knows about Joe, and what is common knowledge at w_1. If, however, v_1 was reached in w_1, then Joe considers it possible that Ida might play O_3 if he chooses I_2. This rationalizes his choice of O_2 at $f(w_1, v_1)$, but contradicts common knowledge of future rationality. More generally, we have the following:

Theorem 8: *Let G be a dynamic game of perfect information*[42] *and take an epistemic model for it. Suppose that in state w current and future rationality is common knowledge at all nodes v. Then* s(w) *is the unique subgame perfect equilibrium profile of G.*

6.3.3 Discussion

While common knowledge of future rationality avoids the difficulties associated with assuming common knowledge of rationality at all decision nodes, it does introduce its own conceptual challenges. The problem is that common knowledge of future rationality embodies a perhaps implausible belief revision policy in dynamic games. The difficulty with exclusively future-oriented reasoning is that it completely discards the past, interpreting any deviations as mistakes that will not be repeated. Indeed, in the game displayed in Figure 18,

[42] Recall, again, that to simplify the exposition, in this section, we are only considering dynamic games with no relevant ties.

at node v_1 Joe knows that Ida did not play her subgame perfect equilibrium strategy. So he knows that her choice at v_0 cannot be explained by common knowledge of rationality. Yet, he refrains from drawing any conclusion from this regarding Ida's future choices. Whatever the explanation of Ida playing I_1, Joe simply ignores it, assuming that Ida made a mistake or perhaps "trembled," and that from now on she comes back to her senses, so to speak, and play rationally at v_2.

It is far from clear whether it is a *rational* belief revision policy to completely discard the past, or to systematically interpret past deviations as one-time mistakes. Perhaps Joe and Ida have played the game in question multiple times, and there is overwhelming evidence of each other's rationality and of each other assuming that the other is rational. In such a context, it might make sense for Joe to discard a single deviation of Ida as a mistake. However, consider a much longer version of the Centipede game from Figure 18, say with 51 decision nodes for Ida, and 50 for Joe. Now, suppose that Joe is at his last choice node. He has observed Ida deviate 50 times from her subgame perfect equilibrium strategy. In other words, none of Ida's past choices can be explained by her being rational and rationality being common knowledge. Does it make sense for Joe to discard all these deviations as mistakes, and assume that, at her very last move, Ida will play rationally? The answer might not be a resounding "no," but it is not a clear "yes" either. It seems that, in cases like this, Joe is failing to take into account strong evidence to the effect that either Ida is not rational, or that some higher-order level of knowledge of rationality fails. More generally, in certain games, common knowledge of future rationality depicts the players as not learning anything from observing what others have done, or not adapting their behavior to these observations (Binmore, 1987; Bicchieri, 1988b). Whatever happens, they fall back on assuming that, from now on, everyone will act as if (future) rationality is common knowledge.

Despite the concerns about the plausibility of common knowledge of future rationality as a belief revision policy, the epistemic explanation of subgame perfect equilibrium represents significant progress. It demonstrates that the limitations of assuming common knowledge of rationality discussed in Section 6.3.1 can be overcome. Common knowledge of future rationality is consistent and results in the players choosing the complete subgame perfect equilibrium profile.

Additionally, this epistemic approach reveals a crucial aspect of subgame perfect equilibrium: It depends on strong assumptions about how players maintain their beliefs about future rational choices despite observing deviations from expected play. This insight opens the door to philosophical discussions about whether such belief revision policies are plausible or rational.

6.4 Extensive Form Rationalizability and Forward Induction

Even though our main focus in this section has been subgame perfect equilibrium, before closing it is worth briefly discussing another type of reasoning for dynamic games, namely *forward induction* reasoning. As with admissibility that we discussed at the end of Section 4, a full presentation of the diverse forms of forward induction reasoning would force us to introduce mathematical details that go beyond the basics that we cover in this Element. Furthermore, as we shall see, the most well-known form of forward induction reasoning and subgame perfect equilibrium only leads to different outcomes or path in games of imperfect information, which we again have bracketed so far in this section. For those reasons, instead of presenting the general details, we only present one example of forward induction reasoning, and provide further references for the reader interested in the underlying theory.

Unlike subgame perfect equilibrium and its associated backward induction algorithm, forward induction is not a clearly defined solution concept for dynamic games. Instead, it functions as an umbrella term for the idea that players infer the strategies and beliefs of others at both current and future decision nodes based on past actions. Backward induction is analogous to mathematical induction, whereas forward induction reflects inductive reasoning in a Humean sense, in which decision-makers draw inferences about future actions from a limited set of past observations.

Examples of forward induction methods include explicable equilibrium (Reny, 1992) and strong Δ-rationalizability, which is a form of rationalizability under constraints (Battigalli, 2003; Battigalli and Siniscalchi, 2003; Catonini, 2019). In this section, we focus on extensive form rationalizability (Pearce, 1984; Battigalli, 2003), which is arguably the most well-studied forward induction method.

The core idea of extensive form rationalizability is that players try, as much as possible, to interpret past choices of the other players as rational. Here, "interpret" means to attribute beliefs to the other players that justify their past choices (Battigalli, 1997). To illustrate, consider the game constructed by Kohlberg (1981), and illustrated in Figure 20. This is a game of imperfect information. Ida initially chooses either to end the game with a payoff of 2 for both players, or to play a static Battle of the Sexes game. Since in that static game the players move simultaneously, this is a dynamic game where neither player observes the other player's move at the second decision node.

The first observation toward forward induction reasoning and extensive form rationalizabilty is that Ida cannot rationally choose I unless she assigns a probability of at least $2/3$ that Joe will choose l. Otherwise, whatever she chooses in

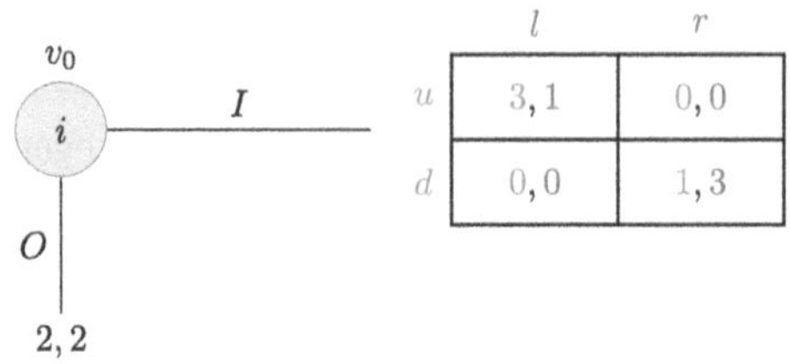

Figure 20 A game where backward and forward induction reasoning lead to different outcomes

the static game, her expected payoff is lower than the guaranteed 2 that she gets from the outside option O. So, if Joe observes Ida choosing I and he believes that this was a rational choice, then he must conclude that Ida believes with probability at least $2/3$ that he will choose l. This is the crux of forward induction reasoning: What Joe does is to attribute *a belief* to Ida, on the basis of which he can maintain his hypothesis that she is rational. In other words, Joe holds his belief in Ida's rationality fixed, as much as possible, and in order to do so he explains her choices by attributing her certain beliefs.

This type of forward induction reasoning has direct consequences for what Ida and Joe would choose in this game. If Joe attributes to Ida a degree of belief of at least 2/3 that he will choose l, then he will also concludes that the only rational choice in the static game for Ida is u. But, of course, if Joe believes that Ida will choose u, his only best response is l. If this reasoning is transparent to both players, then Ida will choose I followed by u and Joe will choose l. By forfeiting a guaranteed payoff of 2, Ida has secured her most preferred outcome (u, l). Observe that a generalized version of subgame perfect equilibrium would not necessarily lead to that outcome. Although (I, d) is strictly dominated, both (O, u) and (O, d) are compatible with subgame perfect equilibrium.

The main epistemic characterization of extensive form rationalizability is in terms of common *strong belief* in rationality (Battigalli and Siniscalchi, 2002). The underlying idea is that a player strongly believes an event E provided she believes E is true at the beginning of the game (in the sense that she assigns probability 1 to E) and continues to believe E as long as it is not ruled out by her evidence. The evidence available to a player in a dynamic game consists of the observations of the previous moves. So, common strong belief in rationality means that the players commonly believe that everyone rationalizes past moves, in the sense of ruling out beliefs under which these moves are not rational.

An important result about backward and forward induction is that in dynamic games of perfect information with no relevant payoff ties, common strong belief in rationality and hence extensive form rationalizability yield the same *outcome* as subgame perfect equilibrium (Battigalli, 1997). This is not to say that the two solution concepts fully coincide: They can make different predictions off the subgame perfect equilibrium path. The underlying idea is that deviations from the subgame perfect equilibrium path may not be rationalizable, and extensive form rationalizability does not impose strong enough constraints in such cases to ensure the full subgame perfect equilibrium profile.[43]

7 Strategic Thinking and Levels of Higher-Order Reasoning in Practice

It should be clear by now that the hierarchies of higher-order belief and knowledge are the cornerstones of epistemic explanations of solution concepts. This naturally raises the question of whether real players, or for that matter anyone engaged in strategic interaction or strategic thinking, base their decisions on that kind of higher-order belief and knowledge. This section addresses these questions by surveying some of the most well-known experimental and empirical results on the Guess the Average game that we discussed in Section 1. The bottom line of these experimental and empirical results is that although they provide rather strong evidence for the fact that we do engage in strategic reasoning involving higher-order beliefs or knowledge, they also suggest that we rarely go beyond the second level.

7.1 Guess the Average in Experiments

The original version of the Guess the Average game, usually called Beauty Contest (Keynes, 1936), was used to explain how the expectations about the beliefs of other agents have a crucial role in the financial markets. Nowadays, the labels "Guess the Average" and "Beauty Contest"[44] encompass a variety of similar games, all having in common the key role played by higher-order strategic reasoning.

The version that has been much used in experiments on higher-order reasoning and that we present here is the same as the one we informally discussed in the introduction, and can be seen as a numerical formulation of Keynes' game due to Moulin (1986). There are N players, each player has to guess an integer

[43] See Heifetz and Perea (2015); Catonini (2019); Perea (2018) for alternative proofs of this result and Arieli and Aumann (2015) for a proof in the special case where each player only moves once.

[44] As mentioned in Section 1, the name "Beauty Contest" goes back to the presentation in Keynes (1936). See Nagel et al. (2017) for an historical overview.

between 0 and 100. The player who guesses closest to two-thirds of the average of all the numbers wins a certain amount. In the case of a tie, the amount is evenly split between the winning players or randomly assigned to one of the winners, depending on the different variants. In the version introduced in Section 1, the prize is evenly split between the winners.

For $N \leq 3$, the only equilibrium of the game – as well as the only rationalizable action profile – is a profile where each player guesses the minimum, zero. When $N > 3$, 0 and 1 are both rationalizable, and both a profile of all zeros and a profile of all ones are equilibria. Choosing 0 or 1, however, may not win the game: If for instance there are more than two players in the game, and a player believes that all the other players are going to pick sufficiently high numbers, then neither 0 nor 1 is a best reply.

7.2 Theories of Mind

Given its structure, the Guess the Average game (in all its possible variants) has been used as a test bed for different theories of the agents' higher-order reasoning in practice. These theories usually go under the name of "theories of minds."[45] Relative to the Guess the Average game, all such theories agree that the number chosen by a player may be revealing of her level in higher-order reasoning, with lower numbers indicating higher levels.

What distinguishes one theory of mind from another is then the way an agent models the other agents in this process of higher-order reasoning. Here, we present the two main theories of mind on the market – the level-*k* reasoning model and the cognitive hierarchy model – and illustrate how each works as well as their differences in a concrete example using the Guess the Average game.

7.2.1 The Level-k Reasoning Model

The first level-*k* reasoning models are due to Nagel (1995) and Stahl and Wilson (1995) and come along with the first experiments on the Guess the Average game. The level-*k* reasoning model, or level-*k* model for short, is based on the concept of iterated best reply (see Section 1), which is used to define the different levels of higher-order reasoning.

First, an agent of level 0 is assumed to have no model of the other agents' beliefs and choices, and to pick an action at random. In the Guess the Average

[45] To be precise, theory of mind is a term generally used in psychology to refer to people's ability to ascribe mental states, such as beliefs, desires, emotions, intentions, and thoughts, to the others (Apperly, 2012). Here, we are particularly interested in the models to ascribe beliefs to the other agents.

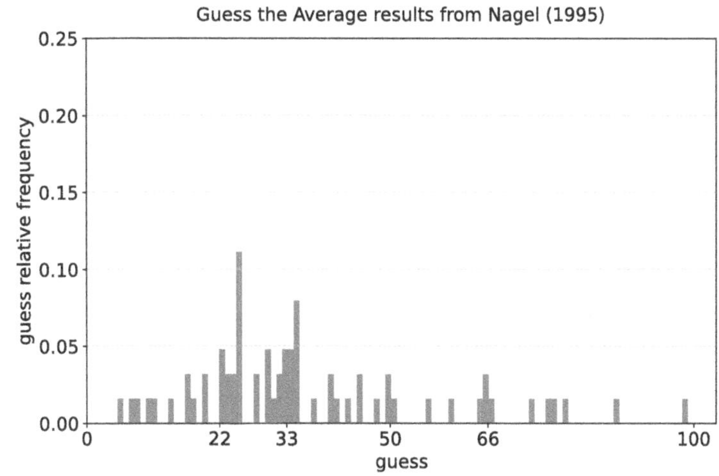

Figure 21 The frequency histogram shows the frequencies of the different guesses obtained by Nagel (1995) for $p = 2/3$

game presented earlier, a level-0 agent hence randomly chooses a number between 0 and 100. Next, an agent of level $k + 1$ believes that the other agents are of level k and, and play her best reply to this belief. In particular, in the Guess the Average game, a level-1 player believes that the others are of level 0 and that they pick a random number. The expectation of a level-1 player is therefore 50, to which the best reply is $2/3 \cdot 50 \approx 33.3$. The level-$k$ model thus predicts that a level-1 agent chooses a number close to 33.3.

By iterating this reasoning, a level-2 agent thinks that the others are of level 1 and that they hence choose numbers close to 33.3, to which she best replies by guessing a number around $(2/3)^2 \cdot 50 \approx 22.2$, and an agent of level $k + 1$ believes that the others are of level k and best replies by guessing a number close to $(2/3)^{k+1} \cdot 50$.[46]

The frequency histogram in Figure 21 shows the results of the lab experiments from Nagel (1995). The x-axis represents the possible guesses of the players divided into 100 bins. The y-axis gives the frequencies of each bin. According to the level-k model, the high frequencies around the numbers 22 and 33 indicate that there is a sizeable amount of players who perform one or

[46] It is important to keep in mind that the solution of Guess the Average we present here, through iterated best reply, follows a different process than the one we informally introduced in Sections 1 and 4, namely rationalizability. In the limit, the two processes make the same prediction, namely that all players choose 0 or 1, depending on the number of players. But rationalizability proceeds by "smaller steps," so to speak, than iterated best reply. Recall that rationalizability only ruled out 100 initially, then 99, and so on. Here, on the other hand, with iterated best reply the players move from a random choice at level 0, to 33 at level 1, 22 at level 2, and so on.

two steps of higher-order reasoning. Spikes corresponding to more than two steps of higher-order reasoning do not appear clearly.

Nagel (1995) conducts three variants of the Guess the Average game, where the target is respectively $1/2$, $2/3$ and $4/3$ of the average guess and observes that most of the guesses are in the neighborhood of $50p$ and $50p^2$, for all three treatments $p \in \{1/2, 2/3, 4/3\}$. Nagel (1995) then concludes that the level-k model can explain such behavior and it can be thus considered a good model for the higher-order reasoning of the players, with the majority of the observations displaying one or two steps of higher-order reasoning.

7.2.2 The Cognitive Hierarchy Model

A famous alternative to the level-k model has been advanced some years later by Camerer et al. (2004), and it is named cognitive hierarchy (CH) model. The central tenet and main difference with respect to the level-k model is that the CH model assumes that an agent of level k does not believe that all the other agents are of level $k-1$, but rather holds a belief represented by a probability distribution over all the lower levels $0, \ldots, k-1$.

In particular, the CH model assumes that the different higher-order levels are distributed according to a Poisson distribution

$$f(k) = \frac{e^{-\lambda}\lambda^k}{k!}$$

where $k = 0, 1, \ldots$ and $f(k)$ is the relative frequency of level-k thinkers in the population. Notice from Figure 22 that the Poisson distribution is a discrete probability distribution over the nonnegative integers, and it is governed by a parameter λ which controls both the mean and the variance of the distribution. Figure 22 shows three examples of Poisson distribution, for $\lambda \in \{1.5, 3, 5\}$. In statistics, the Poisson distribution is generally used to express the probability of an event happening a certain number of times within a given interval of time or space. For instance the number of yellow cars passing in front of your university in the next hour can be modeled as a Poisson random variable. The parameter λ then represents the average number of events occurring within the interval of time or space of interest. From the experiments, Camerer et al. (2004) estimate that the model gives the best fit for $\lambda = 1.5$. This is in line with the intuition that higher-order reasoning levels are hard to perform, and therefore the higher the level the lower the frequency in the population.

One further step is needed to go from the distribution of higher-order levels in the population to the agents' beliefs. We have already mentioned that in the

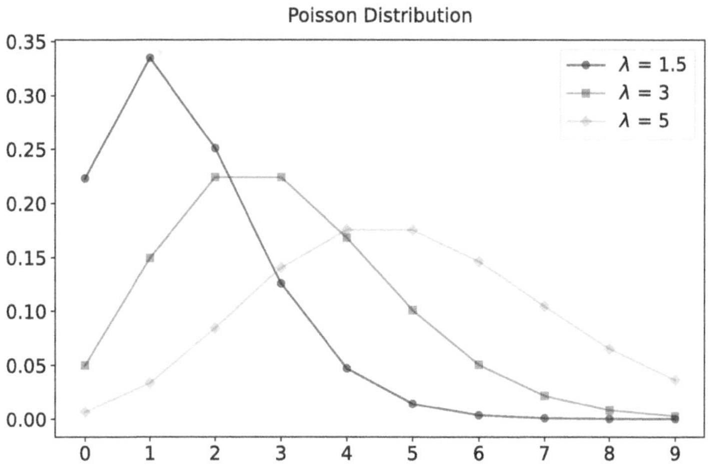

Figure 22 Poisson distribution

CH model an agent of level k believes that the other agents may be of any lower level $0, \ldots, k-1$, while the Poisson distribution is defined over all the nonnegative integers. The belief of an agent of level k is therefore defined by truncating the Poisson distribution at $k-1$: The belief of a level-k agent that another agent is of level h is thus

$$b_k(h) = \begin{cases} \frac{f(h)}{\sum_{l=0}^{k-1} f(l)} & \text{if } h < k \\ 0 & \text{otherwise.} \end{cases}$$

Figure 23 shows some examples of truncated Poisson distribution for $\lambda = 1.5$. The truncated (light-gray) distribution in the top-left picture of Figure 23 represents the belief of a level-2 agent, while the truncated (light-grey) distribution in the bottom-right picture is the belief of a level-5 agent. It is evident from the pictures that the higher the level of an agent, the closer her belief to the actual (dark-grey) distribution of levels in the population, a feature which is called 'increasingly rational expectations' by Camerer et al. (2004).

Finally, as with the level-k model, each agent best replies to her belief. The payoff to agent 1 when she plays action a_1 against action a_2 of agent 2 is denoted as usual by $\pi_1(a_1, a_2)$. The expected payoff of an agent 1 of level k from action a_1 is therefore given by

$$E_k[\pi_1 | a_1] = \sum_{l=0}^{k-1} b_k(l) \pi_i(a_1, a_2^{*l})$$

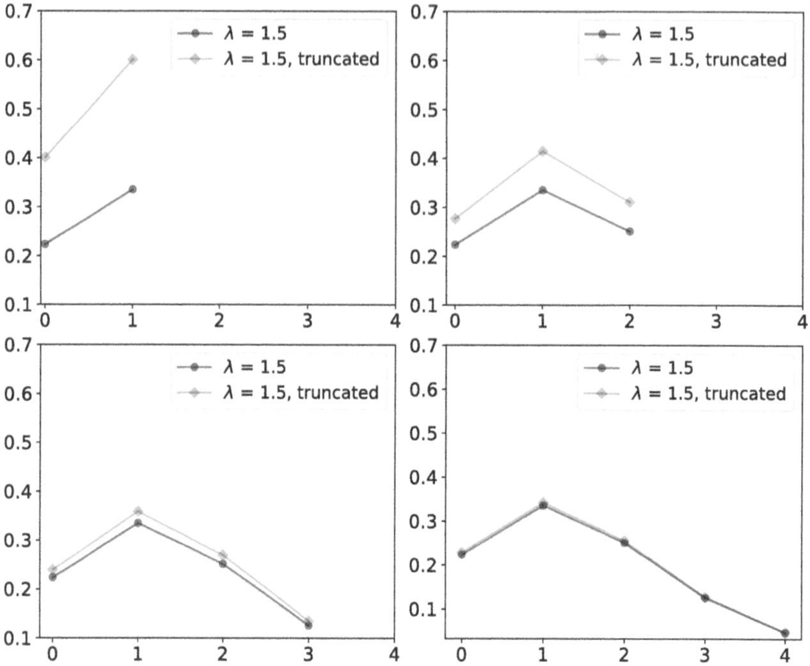

Figure 23 Truncated Poisson distributions

where it is assumed that an agent j of level l always plays a best reply a_j^{*l}. Under the assumption that level-0 players choose an action uniformly at random, the model can then be solved recursively, starting from level-0 players.[47]

For example, in the case of the Guess the Average game the higher-order levels are computed as follows. Level-0 agents choose a number at random, while the truncated Poisson distribution of level-1 agents gives probability 1 to level-0 agents. The expectation of a level-1 agent about the other agents is hence 50, to which her best reply is $2/3 \cdot 50 \approx 33.3$. One level higher, the truncated Poisson distribution (for $\lambda = 1.5$) of a level-2 agent gives probability 0.4 to the others being of level 0 and 0.6 to the others being of level 1. The best reply of a level-2 agent is thus $2/3 \cdot (0.4 \cdot 50 + 0.6 \cdot 33) \approx 26.7$. In the same way, the behavior of all higher-order levels is computed.

A couple of final remarks may be relevant at this point. First, the parameter $\lambda = 1.5$ estimated by Camerer et al. (2004) comes from a multitude of different

47 It is also interesting to notice that because of the way the agents' beliefs are defined in the two models, an agent of level k in the level-k model always respects rationality and mutual belief in rationality of level k (i.e., chooses only actions that survive k iterations of the ρ operator from Section 4), while an agent of level k from the CH model may break rationality and mutual belief in rationality of level k.

variants of the Guess the Average game as well as various other games. If one focuses on the experimental data from Nagel (1995), the best fit is given by $\lambda = 1.1$, according to which level-1 players choose $2/3 \cdot 50 \approx 33.3$ and level-2 players choose $2/3 \cdot (0.48 \cdot 50 + 0.52 \cdot 33) \approx 27.4$.

From this example, it is already evident that in many cases it may not be easy to tell whether the cognitive hierarchy model can explain the data better than the level-k model or vice versa. In general, however, it is an assumption of both models that the agents with different higher-order levels come from an heterogeneous population, where the frequencies of level-0 agents that best fit the data are often zero or close to zero. In this sense, level-0 agents are just a mental construct that higher-level agents use to start their strategic reasoning (Crawford et al., 2013). In some cases, then, the level-k model seems to fit the data better than the CH model, as the latter constraints the different levels to be Poisson distributed with a nonzero percentage of level-0 players (but see also Chong et al. (2005)), while the former does not a priori assume any distribution of the levels in the population and can hence fit the possible absence of level-0 players more easily.

Second, such experiments on higher-order reasoning in general and on the Guess the Average game in particular are conducted with participants that are trained in economic, strategic, or at least quantitative reasoning in the vast majority of the cases. For instance, the findings in Nagel (1995) come from experiments with university students, while the experiments in Camerer et al. (2004) have subject pools consisting of university students, CEOs, or game-theorists. This may raise the question whether these results and models can represent the population at large or are suitable only for highly sophisticated reasoners. We address this point too in the next section.

7.2.3 Comparing the Level-k and CH Models

It is first worth stressing that higher-order interactive reasoning is hard for humans (Kinderman et al., 1998), and this may be an explanation of why almost no level higher than 2 has been observed in the experiments presented earlier. To appreciate the difficulty, consider the following somehow "linear" causal chain: Ida arrived late at work because there was a traffic jam because a dog run in the middle of the street because Kai left the house door open because Joe forgot to lock the door. Human beings usually have no problem parsing and understanding such a "linear" chain, or even much longer ones. Consider instead a higher-order "ping-pong" chain like the following: Ida believes that Joe believes that Kai believes that Ida believes that Joe would go to the football match. Most of us have trouble understanding what's happening after 2–3 levels of interactive beliefs.[48] From an experimental point of view, however, the question of whether the models presented in the last section are portable to other interactive situations and/or generalizable to different subject pools remains.

In general, university students, and students in economics especially, may be particularly trained in abstract thinking and strategic reasoning, and the fact that the experiments presented in the previous section are based on subjects pools of experts or university students may cast some doubts on the generalizability of those results to other subject pools. In this respect, Bosch-Doménech et al. (2002) analyze three different large-scale experiment on the Guess the Average game where participants have been recruited through the German magazine *Spektrum der Wissenschaft*, the British newspaper *Financial Times*, and the Spanish newspaper *Expansión*. Such subject pools are then possibly more diverse than pools consisting only of university students. In the same paper, the authors also present results from experiments with other subject pools (see Figure 24). Looking at Figure 24, the same spikes around the numbers 33 and 22 seem to be present in all the experiments with different subject pools, including the newspaper experiments, which is considered by Bosch-Doménech et al. (2002) as evidence that the level-k model generalizes to subject pools other than university students. However, it is still possible that the similarity in the results between newspaper and other experiments may be due to having chosen markedly economic-oriented newspapers, with reader possibly trained in abstract and strategic reasoning.

There is substantial experimental evidence demonstrating that both the CH model and the level-k model successfully predict behavior across different games. Early foundational studies include Stahl and Wilson (1994), Stahl and Wilson (1995), Costa-Gomes et al. (2001), Costa-Gomes and Crawford (2006), and Crawford and Iriberri (2007). Building on this work, Wright and Leyton-Brown (2017, 2019) show that an extension of the CH model provides the most accurate predictions of human behavior across

[48] This fact that higher-order reasoning seems cognitively very demanding might explain why most experimental results use iterated best reply instead of rationalizability. Recall from footnote 46 that rationalizability proceeds in smaller steps. So to reach the prediction of level-1 reasoners, that is, around 33, rationalizability takes roughly 66 steps. We saw in Section 4, however, that each additional step of rationalizability naturally corresponds to an extra level in the higher-order belief hierarchy. So accounting for the empirical data using rationalizability boils down to assuming that the participants go through roughly 66 or even 78 steps of higher-order reasoning, which seems highly implausible given what we know about the difficulty of such reasoning.

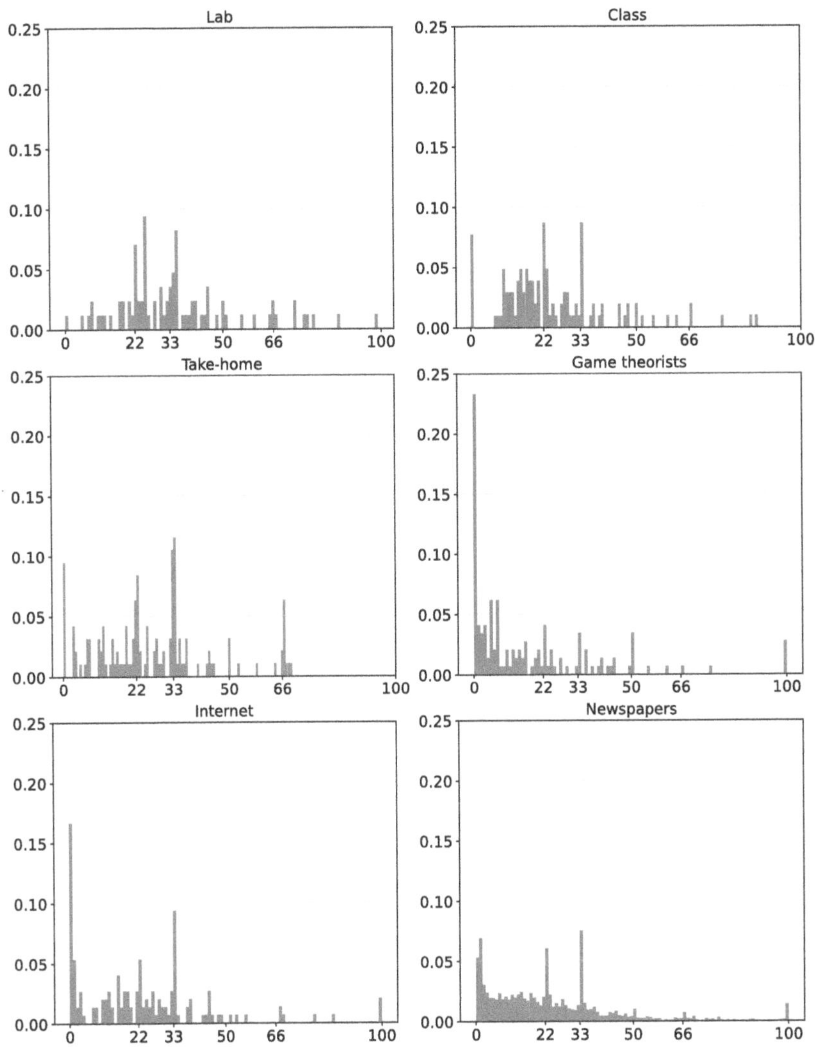

Figure 24 Results from the experiments with different subject pools analyzed by Bosch-Doménech et al. (2002). The lab results come from lab experiments with undergrads from various departments at Bonn University and Caltech. Class and take-home experiments were performed with second-year undergrads in economics at Pompeu Fabra University. Experiments with game theorists come from third- to fourth-year undergraduates in game theory class at Bonn University. Responses for the internet experiments have been collected by email (see Selten and Nagel (1998)). Newspaper experiments aggregate the results from the three experiments mentioned in the main text.

multiple experimental datasets. For a more extensive discussion and comprehensive review of behavioral models in strategic settings, see Crawford et al. (2013) and d'Eon et al. (2024).

7.3 Discussion

The experimental and empirical results that we just surveyed provide two important data points. First, they give evidence for the fact that we, or at least the participants, do engage in some forms of higher-order reasoning. Second, however, this evidence also suggests that we rarely go beyond level 2 in that game, or perhaps level 3 or 4 in other contexts (Kneeland, 2015; Veltman et al., 2019).

Are epistemic explanations "falsified" by those empirical results? We do not think so. While epistemic explanation and level-k reasoning or cognitive hierarchies are both explanatory, they explain different things with different kinds of explanations. Level-k reasoning and cognitive hierarchies are candidates "how-actual" explanations of specific, recurring patterns in empirical data. So their target or *explanandum* is behavioral data, and their *explanans* are the concrete reasoning mechanisms postulated by the level-k or the cognitive hierarchy model. Epistemic characterizations, on the other hand, provide "how-possible" explanations for solution concepts. Their *explanandum* are well-defined sets of actions or strategies in large classes of mathematical models of strategic interaction, that is, classes of games. Their *explanans* are epistemic conditions under which any of those profiles *can* be seen as the result of rational choices.

This doesn't mean, however, that these empirical results and their accompanying theoretical explanations can or even should be ignored by epistemic game theory; quite the contrary. The literature on Guess the Average, as most other empirical work cited earlier, suggests that higher-order reasoning in games is clearly bounded in depth. The modeling tools developed in epistemic game theory and in interactive epistemology can and also have helped us understand what are, in general, the behavioral consequences of bounded higher-order reasoning, for example, (Brandenburger et al., 2021). Analyses such as these make concrete, testable predictions regarding rational behavior in certain classes of games, and this builds a bridge from the general, how-possible explanations typical of traditional epistemic game theory to more how-actual explanations of actual strategic interactions.

8 Conclusion

We presented epistemic game theory as the project of providing foundations, or "epistemic explanations," for game-theoretic solutions. Each section featured prominent instances of such explanations. Nash equilibrium has been characterized in Section 3 in terms of mutual knowledge of rationality and choices. The fundamental theorem (Section 4) established a correspondence between rationalizable actions, essentially actions playable under common belief of rationality, and actions that would survive iterated elimination of strictly dominated strategies. Turning to dynamic games (Section 6), we saw that subgame perfect equilibrium can be characterized in terms of common knowledge of future rationality.

These epistemic explanations have, in turn, shed light on these solution concepts and sparked philosophical debates. The discussion around subgame perfect equilibrium is a good example of that. On the one hand, the epistemic analysis of this solution concept has arguably been motivated by philosophical questions regarding the interpretation of past choices, especially when they contradict the predictions of subgame perfect equilibrium. On the other hand, characterizing subgame perfect equilibrium in terms of common knowledge of future rationality reveals that this solution concept relies on a strict – perhaps overly strict – policy of systematically treating past choices as mistakes or "trembles."

In view of all this, can we then say that epistemic game theory, as a research program, is a success? We think it is, but the answer to this question depends on what counts as success. Epistemic characterizations are not uncontroversial, even among those working in epistemic game theory. The characterization of Nash equilibrium, for instance, is often viewed with skepticism. Our discussions of the epistemic interpretation of mixed strategies, or of common knowledge of future rationality in dynamic games, were also rather critical.

It is easy, however, to overlook the fact that the epistemic explanations themselves enabled these critical discussions. It is, for instance, *because* we know that correctness of belief plays such a central role in the epistemic explanation of Nash equilibria that we can compare the plausibility of the epistemic and the steady-state interpretations of such equilibria. This finding can, in turn, be seen as a way to substantiate many of the conceptual worries raised early on about this equilibria, and motivate the study of other solutions that are based on different, and perhaps philosophically more plausible solutions.

In other words, the goal of epistemic game theory is not, and arguably hasn't been, to justify the use of certain solution concepts, but rather to "make sense" (Spohn, 1982) of them by pinpointing the conditions under which rational players would arrive at these solutions. If these conditions turn out to be controversial, counterintuitive, or implausible, we shouldn't blame the messengers, viz. epistemic characterizations themselves. Whether plausible or not, epistemic characterizations still teach us something about their solutions, and the lessons learned can and, as we saw throughout the Element, have been used to further our understanding of rational decision-making in strategic interactions. Epistemic game theory enables a critical and, perhaps more importantly, *informed* assessment of game-theoretic solutions. This is why we view it as a success.

We could not, of course, cover all the areas of epistemic game theory in this short monograph, even those that have by now given rise to a rich and active literature. Furthermore, we have decided to focus on the most classical results and discussions in epistemic game theory, which might not give a fully accurate impression of the field at the time of writing.

For example, research directions from decision theory and interactive epistemology have naturally been incorporated into epistemic game theory. A recent stream of research in interactive epistemology emphasizes that for an agent to know or believe something, they must first be aware of it. Propositions such as "Ida is aware that Joe believes that Kai is aware of event E" are clearly relevant in strategic contexts, and awareness has been studied as an epistemic attitude crucial to strategic settings (Heifetz et al., 2006; Halpern and Rêgo, 2009).

Another important epistemic attitude, primarily developed in decision theory but not covered here, is *ambiguity*. Also known as unmeasurable or Knightian uncertainty, ambiguity refers to non-probabilistic uncertainty – that is, situations in which an agent's beliefs cannot be represented by a probability distribution over the states of the world. Research on ambiguity in decision theory dates back at least to Ellsberg (1961) and has led to the development of alternative decision criteria for choice under ambiguity, with expected utility maximization typically serving as a special case (Gilboa and Schmeidler, 1989; Schmeidler, 1989; Klibanoff et al., 2005; Hayashi, 2008; Stoye, 2011).

There are at least two ways in which ambiguity can be crucial in game-theoretic contexts. First, a player may be unable to quantify their uncertainty about others' actions in a probabilistic manner. Second, once ambiguity and different decision criteria for choice under uncertainty are introduced, players may also be uncertain about which criteria others are using. In such cases, propositions like "Ida believes that Joe is an expected utility maximizer while Kai is a regret minimizer" become fundamental to the strategic interaction. Consequently, later works in epistemic game theory introduced models that incorporate players with ambiguous beliefs and different decision criteria (Epstein and Wang, 1996; Lo, 1996; Marinacci, 2000; Kajii and Ui, 2005; Ahn, 2007; Di Tillio, 2008; Renou and Schlag, 2010; Halpern and Pass, 2012; Battigalli et al., 2015; Bjorndahl et al., 2017; Galeazzi and Marti, 2023).

Alongside awareness and ambiguity, this Element does not cover several notable topics, including the extensive literature on large or *universal* type spaces (Mertens and Zamir, 1985; Brandenburger and Dekel, 1993), work on learning (Fudenberg and Levine, 1998), and the literature on paradoxes of self-reference in interactive epistemology (Brandenburger and Keisler, 2006; Pacuit, 2007).

As for current research – bearing in mind that such remarks are bound to become quickly outdated – it is fair to say that epistemic game theory has increasingly focused on the study of solution concepts for games of incomplete information, where players face uncertainty about the structure of the game itself (Zuazo-Garin, 2017; Bach and Perea, 2020), epistemic analyses of equilibrium refinements and forward induction (Bach and Cabessa, 2023), and the better integration of empirical findings on higher-order reasoning (Brandenburger et al., 2021).

We conclude with an appeal to fellow philosophers. Many conceptual discussions in epistemic game theory are inherently philosophical, yet philosophers – whether mathematically inclined or not – are less present in the field than they once were. A glance at early epistemic game theory citations reveals seminal contributions by philosophers such as Bicchieri, Skyrms, Stalnaker, and Spohn. Given the prominence of topics like incomplete information and forward induction, the expertise of formal epistemologists, decision theorists, and action theorists is more relevant than ever. We hope this monograph will inspire renewed interest in epistemic game theory within these research communities.

References

Abramsky, S. and Zvesper, J. (2015). From Lawvere to Brandenburger–Keisler: Interactive forms of diagonalization and self-reference. *Journal of Computer and System Sciences*, 81(5):799–812.

Ahn, D. S. (2007). Hierarchies of ambiguous beliefs. *Journal of Economic Theory*, 136(1):286–301.

Alchourrón, C. E., Gärdenfors, P., and Makinson, D. (1985). On the logic of theory change: Partial meet contraction and revision functions. *The Journal of Symbolic Logic*, 50(2):510–530.

Apperly, I. A. (2012). What is "theory of mind"? Concepts, cognitive processes and individual differences. *Quarterly Journal of Experimental Psychology*, 65(5):825–839.

Apt, K. R. (2007). Relative strength of strategy elimination procedures. *Economics Bulletin*, 3(21):1–9.

Apt, K. R. (2011). Direct proofs of order independence. *Economics Bulletin*, 31(1):106–115.

Arieli, I. and Aumann, R. J. (2015). The logic of backward induction. *Journal of Economic Theory*, 159:443–464.

Armbruster, W. and Böge, W. (1979). Bayesian game theory. In Moeschlin, O. and Pallaschke, D., editors, *Game Theory and Related Topics*, pages 17–28. North-Holland.

Aumann, R. and Brandenburger, A. (1995). Epistemic conditions for Nash equilibrium. *Econometrica: Journal of the Econometric Society*, 63(5):1161–1180.

Aumann, R. J. (1974). Subjectivity and correlation in randomized strategies. *Journal of Mathematical Economics*, 1(1):67–96.

Aumann, R. J. (1976). Agreeing to disagree. *The Annals of Statistics*, 4(6):1236–1239.

Aumann, R. J. (1987). Correlated equilibrium as an expression of Bayesian rationality. *Econometrica: Journal of the Econometric Society*, 55(1): 1–18.

Aumann, R. J. (1998). On the centipede game. *Games and Economic Behavior*, 23(1):97–105.

Aumann, R. J. (1999). Interactive epistemology I: Knowledge. *International Journal of Game Theory*, 28(3):263–300.

Aumann, R. J. (2024). *Interactive Epistemology*, volume 10 of *World Scientific Series in Economic Theory*. World Scientific, Singapore.

Aumann, R. J. and Dreze, J. H. (2008). Rational expectations in games. *American Economic Review*, 98(1):72–86.

Aumann, R. J. and Hart, S. (1992). *Handbook of Game Theory with Economic Applications*, volume 2. Elsevier.

Bach, C. W. and Cabessa, J. (2023). Lexicographic agreeing to disagree and perfect equilibrium. *Journal of Mathematical Economics*, 109:102908.

Bach, C. W. and Perea, A. (2020). Generalized Nash equilibrium without common belief in rationality. *Economics Letters*, 186:108526.

Bach, C. W. and Perea, A. (2023). Structure-preserving transformations of epistemic models. *Economic Inquiry*, 61(3):693–719.

Bach, C. W. and Tsakas, E. (2014). Pairwise epistemic conditions for Nash equilibrium. *Games and Economic Behavior*, 85:48–59.

Bacharach, M. (1985). Some extensions of a claim of Aumann in an axiomatic model of knowledge. *Journal of Economic Theory*, 37(1):167–190.

Bacharach, M. (1999). Interactive team reasoning: A contribution to the theory of co-operation. *Research in Economics*, 53(2):117–147.

Bacharach, M. and Mongin, P. (1994). Epistemic logic and the foundations of game theory. *Theory and Decision*, 37:1–6.

Baltag, A., Smets, S., and Zvesper, J. A. (2009). Keep "hoping" for rationality: A solution to the backward induction paradox. *Synthese*, 169(2):301–333.

Barelli, P. (2009). Consistency of beliefs and epistemic conditions for Nash and correlated equilibria. *Games and Economic Behavior*, 67(2):363–375.

Barwise, J. (1988). Three views of common knowledge. In *Proceedings of the 2nd Conference on Theoretical Aspects of Reasoning about Knowledge*, pages 365–379.

Basu, K. (1990). On the non-existence of a rationality definition for extensive games. *International Journal of Game Theory*, 19:33–44.

Basu, K. (1994). The Traveler's dilemma: Paradoxes of rationality in game theory. *The American Economic Review*, 84(2):391–395.

Battigalli, P. (1997). On rationalizability in extensive games. *Journal of Economic Theory*, 74(1):40–61.

Battigalli, P. (2003). Rationalizability in infinite, dynamic games with incomplete information. *Research in Economics*, 57(1):1–38.

Battigalli, P., Cerreia-Vioglio, S., Maccheroni, F., and Marinacci, M. (2015). Self-confirming equilibrium and model uncertainty. *American Economic Review*, 105(2):646–677.

Battigalli, P. and Siniscalchi, M. (1999). Hierarchies of conditional beliefs and interactive epistemology in dynamic games. *Journal of Economic Theory*, 88:188–230.

Battigalli, P. and Siniscalchi, M. (2002). Strong belief and forward induction reasoning. *Journal of Economic Theory*, 106(2):356–391.

Battigalli, P. and Siniscalchi, M. (2003). Rationalization and incomplete information. *Advances in Theoretical Economics*, 3(1):1–43

Bernheim, B. D. (1984). Rationalizable strategic behavior. *Econometrica*, 52(4):1007–1028.

Bicchieri, C. (1988a). Common knowledge and backward induction: A solution to the paradox. In *Proceedings of the 2nd Conference on Theoretical Aspects of Reasoning about Knowledge*, pages 381–393.

Bicchieri, C. (1988b). Strategic behavior and counterfactuals. *Synthese*, 76:135–169.

Bicchieri, C. (1995). The epistemic foundations of Nash equilibrium. In Little, D., editor, *On the Reliability of Economic Models: Essays in the Philosophy of Economics*, pages 91–137. Springer.

Binmore, K. (1987). Modeling rational players: Part I. *Economics & Philosophy*, 3(2):179–214.

Bjorndahl, A., Halpern, J., and Pass, R. (2017). Reasoning about rationality. *Games and Economic Behavior*, 104:146–164.

Bjorndahl, A. and Halpern, J. Y. (2017). From type spaces to probability frames and back, via language. In Lang, J., editor, *TARK 2017 EPTCS 251, Electronic Proceedings in Theoretical Computer Science*, pages 75–87.

Böge, W. and Eisele, T. (1979). On solutions of Bayesian games. *International Journal of Game Theory*, 8:193–215.

Bonanno, G. (1991). The logic of rational play in games of perfect information. *Economics & Philosophy*, 7(1):37–65.

Bonanno, G. (2014). A doxastic behavioral characterization of generalized backward induction. *Games and Economic Behavior*, 88:221–241.

Bonanno, G. (2015). Epistemic foundations of game theory. In van Ditmarsch, H., Halpern, J., van der Hoek, W., and Kooi, B., editors, *Handbook of Logics for Knowledge and Belief*, pages 411–450. College Publications.

Bosch-Doménech, A., Montalvo, J. G., Nagel, R., and Satorra, A. (2002). One, two, (three), infinity, . . .: Newspaper and lab beauty-contest experiments. *The American Economic Review*, 92(5):1687–1701.

Bozdag, S. and Roy, O. (2024). Common knowledge and publicity: Extended abstract. *15th Conference on Logic and the Foundations of Decision and Game Theory (LOFT15)*.

Brandenburger, A. (2010). Origins of epistemic game theory. In Hendricks, V. F. and Roy, O., editors, *Epistemic Logic: Five Questions*, pages 59–69. Automatic Press.

Brandenburger, A. (2014). *The Language of Game Theory: Putting Epistemics into the Mathematics of Games*, volume 5 of *World Scientific Series in Economic Theory*. World Scientific.

Brandenburger, A., Danieli, A., and Friedenberg, A. (2021). The implications of finite-order reasoning. *Theoretical Economics*, 16(4):1605–1654.

Brandenburger, A. and Dekel, E. (1987). Rationalizability and correlated equilibria. *Econometrica: Journal of the Econometric Society*, 55(6):1391–1402.

Brandenburger, A. and Dekel, E. (1993). Hierarchies of beliefs and common knowledge. *Journal of Economic Theory*, 59:189–198.

Brandenburger, A. and Friedenberg, A. (2008). Intrinsic correlation in games. *Journal of Economic Theory*, 141(1):28–67.

Brandenburger, A., Friedenberg, A., and Keisler, H. J. (2008). Admissibility in games. *Econometrica*, 76(2):307–352.

Brandenburger, A., Friedenberg, A., and Keisler, H. J. (2023). The relationship between strong belief and assumption. *Synthese*, 201:175.

Brandenburger, A. and Keisler, H. J. (2006). An impossibility theorem on beliefs in games. *Studia Logica*, 84:211–240.

Caie, M. (2016). Agreement theorems for self-locating belief. *The Review of Symbolic Logic*, 9(2):380–407.

Camerer, C. F., Ho, T.-H., and Chong, J.-K. (2004). A cognitive hierarchy model of games. *The Quarterly Journal of Economics*, 119(3):861–898.

Catonini, E. (2019). Rationalizability and epistemic priority orderings. *Games and Economic Behavior*, 114:101–117.

Catonini, E. and Penta, A. (2022). Backward induction reasoning beyond backward induction.

Chant, S. R. and Ernst, Z. (2008). Epistemic conditions for collective action. *Mind*, 117(467):549–573.

Chen, Y.-C., Long, N. V., and Luo, X. (2007). Iterated strict dominance in general games. *Games and Economic Behavior*, 61(2):299–315.

Chong, J.-K., Camerer, C. F., and Ho, T.-H. (2005). Cognitive hierarchy: A limited thinking theory in games. In Zwick, R. and Rapoport, A., editors, *Experimental Business Research*, pages 203–228, Springer.

Costa-Gomes, M., Crawford, V. P., and Broseta, B. (2001). Cognition and behavior in normal-form games: An experimental study. *Econometrica*, 69(5):1193–1235.

Costa-Gomes, M. A. and Crawford, V. P. (2006). Cognition and behavior in two-person guessing games: An experimental study. *American Economic Review*, 96(5):1737–1768.

Crawford, V. P., Costa-Gomes, M. A., and Iriberri, N. (2013). Structural models of nonequilibrium strategic thinking: Theory, evidence, and applications. *Journal of Economic Literature*, 51(1):5–62.

Crawford, V. P. and Iriberri, N. (2007). Level-k auctions: Can a nonequilibrium model of strategic thinking explain the winner's curse and overbidding in private-value auctions? *Econometrica*, 75(6):1721–1770.

Cubitt, R. P. and Sugden, R. (2003). Common knowledge, salience and convention: A reconstruction of David Lewis' Game theory. *Economics & Philosophy*, 19(2):175–210.

de Bruin, B. (2010). *Explaining Games: The Epistemic Programme in Game Theory*, volume 346 of *Synthese Library*. Springer.

Dekel, E. and Siniscalchi, M. (2015). Epistemic game theory. In *Handbook of Game Theory with Economic Applications*, volume 4, chapter 12, pages 619–702. Elsevier.

d'Eon, G., Greenwood, S., Leyton-Brown, K., and Wright, J. R. (2024). How to evaluate behavioral models. In *Proceedings of the AAAI Conference on Artificial Intelligence*, volume 38, pages 9636–9644.

Di Tillio, A. (2008). Subjective expected utility in games. *Theoretical Economics*, 3(3):619–702.

Dominiak, A. and Lefort, J.-P. (2015). "Agreeing to disagree" type results under ambiguity. *Journal of Mathematical Economics*, 61:119–129.

Ellsberg, D. (1961). Risk, ambiguity, and the Savage axioms. *Quarterly Journal of Economics*, 75:4:643–669.

Epstein, L. and Wang, T. (1996). "Beliefs about beliefs" without probabilities. *Econometrica*, 64(6):1343–1373.

Fagin, R., Halpern, J. Y., Moses, Y., and Vardi, M. Y. (1995). *Reasoning about Knowledge*. The MIT Press.

Fudenberg, D. and Levine, D. K. (1998). *The Theory of Learning in Games*, volume 2. MIT press.

Fukuda, S. (2024). The existence of universal qualitative belief spaces. *Journal of Economic Theory*, 216:287–323.

Galeazzi, P. and Lorini, E. (2016). Epistemic logic meets epistemic game theory: A comparison between multi-agent Kripke models and type spaces. *Synthese*, 193:2097–2127.

Galeazzi, P. and Marti, J. (2023). Choice structures in games. *Games and Economic Behavior*, 140(C):431–455.

Ganguli, J., Heifetz, A., and Lee, B. S. (2016). Universal interactive preferences. *Journal of Economic Theory*, 162:237–260.

Gauthier, D. (1987). *Morals by Agreement*. Clarendon Press.

Gilbert, M. (1992). *On Social Facts*. Princeton University Press.

Gilboa, I. (2009). *Theory of Decision under Uncertainty*. Econometric Society Monographs. Cambridge University Press.

Gilboa, I., Kalai, E., and Zemel, E. (1990). On the order of eliminating dominated strategies. *Operations Research Letters*, 9(2):85–89.

Gilboa, I. and Schmeidler, D. (1989). Maxmin expected utility with non-unique prior. *Journal of Mathematical Economics*, 18:141–153.

Gintis, H. (2009). The local best response criterion: An epistemic approach to equilibrium refinement. *Journal of Economic Behavior & Organization*, 71(2):89–97.

Grosskopf, B. and Nagel, R. (2008). The two-person beauty contest. *Games and Economic Behavior*, 62(1):93–99.

Halpern, J. and Moses, Y. (2017). Characterizing solution concepts in terms of common knowledge of rationality. *International Journal of Game Theory*, 46:457–473.

Halpern, J. Y. (2001). Substantive rationality and backward induction. *Games and Economic Behavior*, 37(2):425–435.

Halpern, J. Y. (2010). Lexicographic probability, conditional probability, and nonstandard probability. *Games and Economic Behavior*, 68(1):155–179.

Halpern, J. Y. and Pass, R. (2009). A logical characterization of iterated admissibility. In *Proceedings of the 12th Conference on Theoretical Aspects of Rationality and Knowledge*, pages 146–155.

Halpern, J. Y. and Pass, R. (2012). Iterated regret minimization: A new solution concept. *Games and Economic Behavior*, 74(1):194–207.

Halpern, J. Y. and Rêgo, L. C. (2009). Reasoning about knowledge of unawareness. *Games and Economic Behavior*, 67(2):503–525.

Harsanyi, J. C. (1967). Games with incomplete information played by "Bayesian" players, I–III part I. The basic model. *Management Science*, 14(3):159–182.

Harsanyi, J. C. (1973). Games with randomly disturbed payoffs: A new rationale for mixed-strategy equilibrium points. *International Journal of Game Theory*, 2(1):1–23.

Harsanyi, J. C. and Selten, R. (1988). A general theory of equilibrium selection in games. *MIT Press Books*, 1.

Hayashi, T. (2008). Regret aversion and opportunity dependence. *Journal of Economic Theory*, 139(1):242–268.

Heifetz, A., Meier, M., and Schipper, B. C. (2006). Interactive unawareness. *Journal of Economic Theory*, 130(1):78–94.

Heifetz, A. and Perea, A. (2015). On the outcome equivalence of backward induction and extensive form rationalizability. *International Journal of Game Theory*, 44:37–59.

Heifetz, A. and Samet, D. (1998a). Knowledge spaces with arbitrarily high rank. *Games and Economic Behavior*, 22(2):260–273.

Heifetz, A. and Samet, D. (1998b). Topology-free typology of beliefs. *Journal of Economic Theory*, 82(2):324–341.

Holt, C. A. and Roth, A. E. (2004). The Nash equilibrium: A perspective. *Proceedings of the National Academy of Sciences of the United States of America*, 101(12):3999–4002.

Huber, F. (2009). Belief and degrees of belief. In Huber, F. and Schmidt-Petri, C., editors, *Degrees of Belief*, pages 1–33. Springer.

Icard, T. (2021). Why be random? *Mind*, 130(517):111–139.

Kajii, A. and Ui, T. (2005). Incomplete information games with multiple priors. *The Japanese Economic Review*, 56:332–351.

Keynes, J. M. (1936). *The General Theory of Employment, Interest and Money*. Macmillan. 14th ed, 1973.

Kinderman, P., Dunbar, R., and Bentall, R. P. (1998). Theory-of-mind deficits and causal attributions. *British Journal of Psychology*, 89(2):191–204.

Klibanoff, P., Marinacci, M., and Mukerji, S. (2005). A smooth model of decision making under ambiguity. *Econometrica*, 73(6):1849–1892.

Kneeland, T. (2015). Identifying higher-order rationality. *Econometrica*, 83(5):2065–2079.

Kohlberg, E. (1981). Some problems with the concept of perfect equilibrium, rapp. rep. NBER Conf. Theory Gen. Econ. Equilibr. K. Dunz N. Singh, University of California, Berkeley.

Kreps, D. and Wilson, R. (1982). Sequential Equilibria. *Econometrica*, 50: 863–894.

Kuechle, G. (2009). What happened to the three-legged centipede game? *Journal of Economic Surveys*, 23(3):562–585.

Kuhn, S. (2024). Prisoner's dilemma. In Zalta, E. N. and Nodelman, U., editors, *The Stanford Encyclopedia of Philosophy*. Metaphysics Research Lab, Stanford University, Winter 2024 edition.

Lederman, H. (2015). People with common priors can agree to disagree. *The Review of Symbolic Logic*, 8(1):11–45.

Leitgeb, H. (2017). *The Stability of Belief: How Rational Belief Coheres with Probability*. Oxford University Press.

Lewis, D. (1969). *Convention: A Philosophical Study*. John Wiley & Sons.

Lin, H. and Kelly, K. (2012). Propositional reasoning that tracks probabilistic reasoning. *Journal of Philosophical Logic*, 41(6):957–981.

Lipman, B. L. (1995). Comments on Bicchieri: The epistemic foundations of Nash equilibrium. In Little, D., editor, *On the Reliability of Economic Models: Essays in the Philosophy of Economics*, pages 138–146. Springer.

Lo, K. C. (1996). Equilibrium in beliefs under uncertainty. *Journal of Economic Theory*, 71(2):443–484.

Marinacci, M. (2000). Ambiguous games. *Games and Economic Behavior*, 31(2):191–219.

McLennan, A. (1985). Justifiable beliefs in sequential equilibrium. *Econometrica: Journal of the Econometric Society*, 53(4):889–904.

Mertens, J.-F. and Zamir, S. (1985). Formulation of Bayesian analysis for games with incomplete information. *International Journal of Game Theory*, 14:1–29.

Morgenstern, O. (1935). Vollkommene voraussicht und wirtschaftliches gleichgewicht. *Zeitschrift für Nationalökonomie/Journal of Economics*, 6(3):337–357.

Morris, S. (1995). The common prior assumption in economic theory. *Economics & Philosophy*, 11(2):227–253.

Moss, L. S. and Viglizzo, I. D. (2004). Harsanyi type spaces and final coalgebras constructed from satisfied theories. *Electronic Notes in Theoretical Computer Science*, 106:279–295.

Moulin, H. (1986). *Game Theory for Social Sciences*. New York University Press.

Myerson, R. B. (1978). Refinements of the Nash equilibrium concept. *International Journal of Game Theory*, 7:73–80.

Myerson, R. B. (1991). *Game Theory: Analysis of Conflict*. Harvard University Press.

Nagel, R. (1995). Unraveling in guessing games: An experimental study. *The American Economic Review*, 85(5):1313–1326.

Nagel, R., Bühren, C., and Frank, B. (2017). Inspired and inspiring: Hervé moulin and the discovery of the beauty contest game. *Mathematical Social Sciences*, 90:191–207.

Nash, J. F. (1950). Equilibrium points in n-person games. *Proceedings of the National Academy of Sciences of the United States of America*, 36:48–49.

Osborne, M. J. and Rubinstein, A. (1994). *A Course in Game Theory*. MIT Press.

Pacuit, E. (2007). Understanding the Brandenburger-Keisler paradox. *Studia Logica*, 86:435–454.

Palacios-Huerta, I. (2014). *Beautiful Game Theory: How Soccer Can Help Economics*. Princeton University Press.

Paternotte, C. (2011). Being realistic about common knowledge: A Lewisian approach. *Synthese*, 183(2):249–276.

Pearce, D. G. (1984). Rationalizable strategic behavior and the problem of perfection. *Econometrica*, 52(4):1029–1050.

Perea, A. (2007a). Epistemic foundations for backward induction: An overview. In *Interactive logic. Proceedings of the 7th Augustus de Morgan Workshop*, volume 1, pages 159–193. Amsterdam University Press Amsterdam.

Perea, A. (2007b). A one-person doxastic characterization of Nash strategies. *Synthese*, 158(2):251–271.

Perea, A. (2012). *Epistemic Game Theory: Reasoning and Choice*. Cambridge University Press.

Perea, A. (2014a). Belief in the opponents future rationality. *Games and Economic Behavior*, 83:231–254.

Perea, A. (2014b). From classical to epistemic game theory. *International Game Theory Review*, 16(1):1–22.

Perea, A. (2018). Why forward induction leads to the backward induction outcome: A new proof for Battigalli's theorem. *Games and Economic Behavior*, 110:120–138.

Peterson, M. (2009). *An Introduction to Decision Theory*. Cambridge University Press.

Renou, L. and Schlag, K. H. (2010). Minimax regret and strategic uncertainty. *Journal of Economic Theory*, 145(1):264–286.

Reny, P. J. (1988). Common knowledge and games with perfect information. In *PSA: Proceedings of the Biennial Meeting of the Philosophy of Science Association*, volume 1988, pages 363–369. Philosophy of Science Association.

Reny, P. J. (1992). Backward induction, normal form perfection and explicable equilibria. *Econometrica: Journal of the Econometric Society*, 60(3): 627–649.

Reny, P. J. (1993). Common belief and the theory of games with perfect information. *Journal of Economic Theory*, 59(2):257–274.

Resnik, M. D. (1987). *Choices: An Introduction to Decision Theory*. University of Minnesota Press.

Rich, P. (2015). Rethinking common belief, revision, and backward induction. *Mathematical Social Sciences*, 75:102–114.

Rosenthal, R. W. (1981). Games of perfect information, predatory pricing and the chain-store paradox. *Journal of Economic theory*, 25(1):92–100.

Rubinstein, A. (1991). Comments on the interpretation of game theory. *Econometrica*, 59(4):909–924.

Samet, D. (1996). Hypothetical knowledge and games with perfect information. *Games and Economic Behavior*, 17(2):230–251.

Samet, D. (2010). Agreeing to disagree: The non-probabilistic case. *Games and Economic Behavior*, 69(1):169–174.

Samet, D. (2022). The impossibility of agreeing to disagree: An extension of the sure-thing principle. *Games and Economic Behavior*, 132:390–399.

Samuelson, L. (1992). Dominated strategies and common knowledge. *Games and Economic Behavior*, 4(2):284–313.

Schelling, T. C. (1960). *The Strategy of Conflict: With a New Preface by the Author*. Harvard University Press.

Schmeidler, D. (1989). Subjective probability and expected utility without additivity. *Econometrica*, 57:571–589.

Seel, C. and Tsakas, E. (2017). Rationalizability and Nash equilibria in guessing games. *Games and Economic Behavior*, 106(C):75–88.

Selten, R. (1975). Reexamination of the perfectness concept for equilibrium points in extensive games. *International Journal of Game Theory*, 4:1–31.

Selten, R. and Nagel, R. (1998). Das zahlenwahlspiel-hintergründe und ergebnisse. *Spektrum der Wissenschaft, February*, pages 16–22.

Skyrms, B. (1990). *The Dynamics of Rational Deliberation*. Harvard University Press.

Spohn, W. (1982). How to make sense of game theory. In Stegmüller, W., Balzer, W., and Spohn, W., editors, *Philosophy of Economics*, volume 2 of *Studies in Contemporary Economics*, pages 239–270. Springer.

Stahl, D. O. and Wilson, P. (1995). On players' models of other players: Theory and experimental evidence. *Games and Economic Behavior*, 10(1): 218–254.

Stahl, D. O. and Wilson, P. W. (1994). Experimental evidence on players' models of other players. *Journal of Economic Behavior & Organization*, 25(3):309–327.

Stalnaker, R. (1996). Knowledge, belief and counterfactual reasoning in games. *Economics & Philosophy*, 12(2):133–163.

Stalnaker, R. (1999). Extensive and strategic forms: Games and models for games. *Research in Economics*, 53(3):293–319.

Stoye, J. (2011). Axioms for minimax regret choice correspondences. *Journal of Economic Theory*, 146:2226–2251.

Tan, T. C.-C. and da Costa Werlang, S. R. (1988). The Bayesian foundations of solution concepts of games. *Journal of Economic Theory*, 45(2):370–391.

van Benthem, J. (2011). *Logical Dynamics of Information and Interaction*. Cambridge University Press.

van Benthem, J. and Klein, D. (2023). Logics for analyzing games. In Zalta, E. N. and Nodelman, U., editors, *The Stanford Encyclopedia of Philosophy*. Metaphysics Research Lab, Stanford University, Fall 2023 edition.

Van Damme, E. (1983). *Refinements of the Nash Equilibrium Concept*. Lecture Notes in Economics and Mathematical Systems. Springer Berlin Heidelberg.

Van Damme, E. (1984). A relation between perfect equilibria in extensive form games and proper equilibria in normal form games. *International Journal of Game Theory*, 13:1–13.

Van Damme, E. (1991). *Stability and perfection of Nash equilibria*, volume 339. Springer.

van Ditmarsch, H., Halpern, J. Y., van der Hoek, W., Barteld, and Kooi, editors (2015). *Handbook of Epistemic Logic*. College Publications.

Vanderschraaf, P. and Sillari, G. (2023). Common knowledge. In Zalta, E. N. and Nodelman, U., editors, *The Stanford Encyclopedia of Philosophy*. Metaphysics Research Lab, Stanford University, Winter 2023 edition.

Veltman, K., de Weerd, H., and Verbrugge, R. (2019). Training the use of theory of mind using artificial agents. *Journal on Multimodal User Interfaces*, 13: 3–18.

Volij, O. (1996). Epistemic conditions for equilibrium in beliefs without independence. *Journal of Economic Theory*, 70(2):391–406.

von Neumann, J. and Morgenstern, O. (1944). *Theory of Games and Economic Behavior*. Princeton University Press.

Vromen, H. (2023). Reasoning with reasons: Lewis on common knowledge. *Economics & Philosophy*, 40(2):1–22.

Wagenaar, W. A. (1972). Generation of random sequences by human subjects: A critical survey of literature. *Psychological Bulletin*, 77(1):65.

Weibull, J. W. (1997). *Evolutionary Game Theory*. MIT Press.

Werlang, S. (1986). *Common Knowledge and Game Theory*. PhD thesis, Princeton University.

Wright, J. R. and Leyton-Brown, K. (2017). Predicting human behavior in unrepeated, simultaneous-move games. *Games and Economic Behavior*, 106:16–37.

Wright, J. R. and Leyton-Brown, K. (2019). Models of level-0 behavior for predicting human behavior in games. *Journal of Artificial Intelligence Research*, 64:357–383.

Ye, R., Zhao, D., Zhang, M., and Liu, W. (2023). Nash equilibrium and tennis serve performance: A game theory analysis. *International Journal of Performance Analysis in Sport*, 23(6):515–526.

Zollman, K. J. (2022). On the normative status of mixed strategies. In Augustin, T., Cozman, F.G., and Wheeler, G., editors, *Reflections on the Foundations of Probability and Statistics: Essays in Honor of Teddy Seidenfeld*, pages 207–239. Springer.

Zuazo-Garin, P. (2017). Uncertain information structures and backward induction. *Journal of Mathematical Economics*, 71:135–151.

Cambridge Elements

Decision Theory and Philosophy

Martin Peterson
Texas A&M University

Martin Peterson is Professor of Philosophy and Sue and Harry E. Bovay Professor of the History and Ethics of Professional Engineering at Texas A&M University. He is the author of four books and one edited collection, as well as many articles on decision theory, ethics and philosophy of science.

About the Series

This Cambridge Elements series offers an extensive overview of decision theory in its many and varied forms. Distinguished authors provide an up-to-date summary of the results of current research in their fields and give their own take on what they believe are the most significant debates influencing research, drawing original conclusions.

Cambridge Elements

Decision Theory and Philosophy

Elements in the Series

Beyond Uncertainty: Reasoning with Unknown Possibilities
Katie Steele and H. Orri Stefánsson

Evidential Decision Theory
Arif Ahmed

Commitment and Resoluteness in Rational Choice
Chrisoula Andreou

Money-Pump Arguments
Johan E. Gustafsson

Bargaining Theory
Peter Vanderschraaf

Evolutionary Game Theory
J. McKenzie Alexander

The History and Methodology of Expected Utility
Ivan Moscati

Preference Change
David Strohmaier and Michael Messerli

The Measurement of Subjective Probability
Edward J. R. Elliott

Rationality and Time Bias
Abelard Podgorski

Opinion Pooling
Lee Elkin and Richard Pettigrew

Epistemic Game Theory
Paolo Galeazzi, Eric Pacuit and Olivier Roy

A full series listing is available at: www.cambridge.org/EDTP

Printed in the United Kingdom by TJ Clays Ltd.